# Concept Mapping and Education

# Concept Mapping and Education

Editor

**AMM Sharif Ullah**

MDPI • Basel • Beijing • Wuhan • Barcelona • Belgrade • Manchester • Tokyo • Cluj • Tianjin

*Editor*
AMM Sharif Ullah
Faculty of Engineering,
Kitami Institute of Technology
Japan

*Editorial Office*
MDPI
St. Alban-Anlage 66
4052 Basel, Switzerland

This is a reprint of articles from the Special Issue published online in the open access journal *Education Sciences* (ISSN 2227-7102) (available at: https://www.mdpi.com/journal/education/special_issues/Concept_Mapping_and_Education).

For citation purposes, cite each article independently as indicated on the article page online and as indicated below:

LastName, A.A.; LastName, B.B.; LastName, C.C. Article Title. *Journal Name* **Year**, *Article Number*, Page Range.

**ISBN 978-3-03943-392-6 (Hbk)**
**ISBN 978-3-03943-393-3 (PDF)**

# Contents

# About the Editor

**AMM Sharif Ullah** is Chair of the Intelligent Machines and Biomechanics Program and Director of the Advanced Manufacturing Engineering Laboratory at the Kitami Institute of Technology. He is an active researcher in the areas of engineering design and intelligent manufacturing. He received his first degree, Bachelor of Science in Engineering (Mechanical), from the Bangladesh University of Engineering and Technology in 1992. He received his Master's and Doctoral Degrees from Kansai University in 1996 and 1999, respectively. Before joining his current employer in October 2009, he worked as a full-time faculty member at the Asian Institute of Technology (Assistant Professor, 2000–2002) and the United Arab Emirates University (Assistant Professor, 2002–2006; Associate Professor, 2006–2009). He has mentored more than 120 undergraduate/graduate students coming from different countries. He researches knowledge-based systems for product realization, emphasizing creativity, design, manufacturing, operations, materials, sustainability, and systems. He published more than 120 technical articles in reputed peer-reviewed journals, edited books, and international conference proceedings. He serves on the editorial boards of several peer-reviewed international journals. His current research focuses on Industry 4.0, 3D printing, engineering design, sustainable product development, reverse engineering, precision manufacturing, and engineering education.

*Editorial*

# Concept Map and Knowledge

**AMM Sharif Ullah**

Division of Mechanical and Electrical Engineering, Kitami Institute of Technology, 165 Koen-cho, Kitami 090-8507, Japan; ullah@mail.kitami-it.ac.jp; Tel./Fax: +81-157-26-9207

Received: 14 September 2020; Accepted: 14 September 2020; Published: 15 September 2020

---

Based on Piaget's genetic epistemology, Ausubel developed the assimilation theory of verbal learning [1,2]. This theory is considered the foundation of meaningful learning. The outcomes of meaningful learning take the form of concept maps—networks of some selected linguistic expressions and concepts [3]. Thus, concept map-based education helps avoid rote learning. Additionally, it helps prepare content for effective on-ground and e-learning. Moreover, it helps measure learning outcomes at the course, program, and institutional levels. As a result, concept map-based education has been used at school, college, university, and professional levels.

This Special Issue on Concept Mapping and Education solicited manuscripts on the theoretical foundations of concept mapping and its application areas, including (but not limited to) concept mapping in school, higher and professional education; concept mapping in active learning, problem-based learning, and project-based learning; and distance learning. Finally, this Special Issue published five selected articles, each of which went through a rigorous peer-review process. The editor would like to acknowledge the authors and reviewers for their invaluable contributions that enrich concept map-based education in particular and education science in general.

"A Concept Tree of Accounting Theory: (Re)Design for the Curriculum Development" [4] contributes to accounting and concept mapping literature by depicting a concept tree based on the Accounting Theory curriculum. The proposed tree-shaped concept map graphically interprets the sophisticated accounting theories and concepts and their complex interrelationships. In teaching practices, this concept tree promotes curriculum development, systematizing relevant topics, and exam-taking.

"Uncovering Types of Knowledge in Concept Maps" [5] uncovers the types of knowledge in concept maps. Concept maps represent different types of knowledge (e.g., procedural and conceptual). The authors used Legitimation Code Theory (LCT) to analyze concept maps in semantic gravity and semantic density. It was found that different types of knowledge are considered necessary to achieve professional knowledge or expert understanding. The author used examples to demonstrate students' learning patterns towards gaining expert knowledge. The authors described the implication of the study for curriculum design and teaching evaluation.

"The Salutogenic Management of Pedagogic Frailty: A Case of Educational Theory Development Using Concept Mapping" [6] reinterprets "pedagogic frailty" using concept maps, focusing on the salutogenesis of teachers such as assets, wellness, and sense of coherence. The concept maps define the complex relationship between pedagogic frailty and salutogenesis. This study paves the way to an educational theory for enhancing university-wide mental health literacy and avoiding the misapplication of existing teaching quality enhancement models. As a result, it has a greater utility for university managers.

"The Role and Efficacy of Creative Imagination in Concept Formation: A Study of Variables for Children in Primary School" [7] is a contribution towards creative concept formation, which is one of the main concerns of assimilation theory-based concept mapping. The 8–12-year-old school children were tested using four subtests. Three of the subtests were designed to evaluate narrative (verbal) creativity, and the other was designed to evaluate drawing (i.e., graphic) creativity. The tests indicate the child's learning ability regarding fluency, flexibility, originality in narrative representations, and graphics. The study can be used to evaluate concept map-based learning, focusing on creative thinking.

"Fundamental Issues of Concept Mapping Relevant to Discipline-Based Education: A Perspective of Manufacturing Engineering" [8] addresses some fundamental concept mapping issues relevant to discipline-based education (e.g., manufacturing education). This article argues that knowledge-type-aware concept mapping is a solution to create and analyze the semantic web-embedded dynamic knowledge bases for both human and machine learning from the perspective of Industry 4.0. Accordingly, this article defines five types of knowledge, namely, analytic a priori knowledge, synthetic a priori knowledge, synthetic a posteriori knowledge, meaningful knowledge, and skeptic knowledge. These types of knowledge help find some rules and guidelines to create and analyze concept maps for human and machine learning. The presence of these types of knowledge is elucidated using a real-life manufacturing knowledge representation case.

In sum, a problem cannot be solved without applying knowledge. Therefore, educators embark on how to disseminate knowledge to students. At the same time, students embark on how to learn relevant knowledge. Both knowledge dissemination and learning can be carryout using appropriately designed concept maps. In this case, the concept maps must manage the myriad proximal and distal relationships between knowledge and other relevant entities (human/machine learning, logical inferences, experimental data, analytical results, creative thinking, and cognitive reflections). As a result, knowledge-type-aware concept mapping is an effective means of education and learning [9]. Let us cherish this notion.

**Funding:** This research received no external funding.

**Conflicts of Interest:** The author declares no conflict of interest.

## References

1. Ausubel, D.; Novak, J.; Hanesian, H. *Educational Psychology: A Cognitive View*, 2nd ed.; Holt, Rinehart & Winston: New York, NY, USA, 1978.
2. Seel, N.M. Assimilation theory of learning. In *Encyclopedia of the Sciences of Learning*; Seel, N.M., Ed.; Springer: Boston, MA, USA, 2012. [CrossRef]
3. Novak, J.D. *Learning, Creating, and Using Knowledge: Concept Maps as Facilitative Tools in Schools and Corporations*, 2nd ed.; Rutledge: New York, NY, USA, 2012; p. 105644.
4. Zhou, Y. A Concept Tree of Accounting Theory: (Re)Design for the Curriculum Development. *Educ. Sci.* **2019**, *9*, 111. [CrossRef]
5. Kinchin, I.M.; Möllits, A.; Reiska, P. Uncovering Types of Knowledge in Concept Maps. *Educ. Sci.* **2019**, *9*, 131. [CrossRef]
6. Kinchin, I.M. The Salutogenic Management of Pedagogic Frailty: A Case of Educational Theory Development Using Concept Mapping. *Educ. Sci.* **2019**, *9*, 157. [CrossRef]
7. Gonzalez Garcia, J.; Mukhopadhyay, T.P. The Role and Efficacy of Creative Imagination in Concept Formation: A Study of Variables for Children in Primary School. *Educ. Sci.* **2019**, *9*, 175.
8. Ullah, A.S. Fundamental Issues of Concept Mapping Relevant to Discipline-Based Education: A Perspective of Manufacturing Engineering. *Educ. Sci.* **2019**, *9*, 228. [CrossRef]
9. Ullah, A.S. What is knowledge in Industry 4.0? *Eng. Rep.* **2020**, *2*, e12217. [CrossRef]

*Article*

# A Concept Tree of Accounting Theory: (Re)Design for the Curriculum Development

**Yining Zhou**

Sydney City School of Business, Top Education Institute, Eveleigh, NSW 2015, Australia; yining.zhou@top.edu.au

Received: 10 March 2019; Accepted: 16 May 2019; Published: 21 May 2019

**Abstract:** This study contributes to both accounting and concept mapping literature through the depiction of a concept tree based on the Accounting Theory curriculum, which has undergone recent and rapid expansion of its knowledge and has hence outgrown the previous limited mapping work. This tree-shaped concept map not only accounts for a particular mapping approach little studied and scarcely exemplified in literature, but also signifies a creative model that graphically interprets the sophisticated system of accounting theories and concepts as well as their complex interrelationships. In teaching practices, this concept tree has attested a potential to promote curriculum development, as evidenced in sequence and cohesion of topics and by being linked meaningfully to exam design.

**Keywords:** concept map; concept tree; accounting theory; accounting education; curriculum development

## 1. Introduction

Concept maps have been proven to be a useful educational technique, integrating semantic relationships between concepts into node–link diagrams [1,2]. A traditional or Novakian concept map is made up of nodes of concepts, links, and linking phrases that articulate how these concept nodes relate to each other [3]. Through the provision of visual and abstract evident graphic depictions, concept maps can be used to aid organization, preparation, and presentation of information concerning a topic, a lecture, a course, or even a discipline of knowledge [2]. In higher education, concept mapping has been well regarded as a powerful and versatile tool to facilitate learning (particularly beneficial for lower performance students) [4,5], to enhance learners' problem-solving abilities [6,7], to foster knowledge sharing and social artefacts between students [7], to further teacher–learner dialogue [8], and to renew assessment modes from rote recall of isolated knowledge pieces towards connections between ideas [9].

Given its extensive use in a variety of (particularly science-based) disciplines, concept mapping has not received much attention in accounting education [10,11]. Of particular concern is its limited application in the Accounting Theory (and Application) course, where only one curriculum concept map (Simon 2007) has been found from literature thus far [4]. The essential problem is a map-knowledge mismatch. Proposed more than one decade ago and laid on a traditional financial position, Simon's map is unable to sufficiently and appropriately account for the current state of curriculum knowledge, as the system has been profoundly reshaped by theorizing accountability that triggers emergence and development of non-financial accountings [12–15]. Another inadequacy of this map is that it portrays only simple hierarchical relationships used to distinguish between "more general" and "more specific" concepts and is thus unable to make references to curriculum themes such as accounting evolution and sector-specificity, which are required to address the connections and the interconnections among accounting theories and models across different categories. It has been highlighted that concept maps can serve as powerful tools for educators and students to tackle the challenge emerging from the increasing amount and complexity of knowledge, where more sophisticated and alternative thoughts to the existing understanding are continuously generated [5,9]. In this regard, concept maps enable a "knowledge integration" process towards a higher level of knowledge, through which the connections between the old and the new ideas are rigorously explored, and the repertoire of all these inconsistent ideas is systematically rearranged [9]. From this perspective, it is necessary to update the curriculum concept map specified on both the extended, nonfinancial knowledge domain of accounting alternatives to traditional financial accounting [1,16], and the intricate (inter)relationships within.

On the other hand, a particular form of concept maps is to symbolize the system of knowledge or information in a tree structure [17,18]. The presentational effectiveness of tree models is highly praised in such a knowledge system where an evolutionary strain is implicated among concepts, or they are distinctively diversified in contexts of meanings or fields of applications [19,20]. This characteristic inherent to the tree model provides the opportunity to redesign the concept map for the Accounting Theory curriculum, where the financial and the non-financial accounting contexts have been distinctly differentiated but still share the same root of accountability.

The purpose of this paper is to contribute an instructor-prepared tree-like concept map stipulated on the Accounting Theory curriculum. Two insights are concluded from this study. First, the concept tree stands for a creative model that visualizes the sophisticated system of accounting theories and concepts, as well as their complex interrelationships. Second, this tree has proven its potential to promote curriculum development, as evidenced by the sequence and cohesion of topics and a meaningful linkage to exam design. An additional contribution made by this paper is the enrichment of the concept mapping methodology. In doing so, this paper constructs a concrete account to explore the tree-shape as a particular mapping technique that has been rarely studied and inadequately exemplified in literature.

## 2. Concept Map in Accounting Education and a Critique of the Old Curriculum Map

The theoretical ground of concept mapping is traced back to Ausubel's meaningful learning theory [21,22], which is based on the idea that memory-testing or rote learning cannot achieve an efficient learning process. In Ausubel's view, a successful learning process requires a platform or scaffolding through which new knowledge can be absorbed and integrated to the previously-attained knowledge. This platform or scaffolding comes from the notion of "advance organizers", which refers to a knowledge construction method where general or abstract ideas work to depict an overview to assimilate new knowledge, and fragments of texts are reasonably linked [23,24].

Ausulbel's meaningful learning theory [21–23] provides critical insights for Novak's groundwork for the development of concept mapping [1,24], which has been versatile and flexible in its application for a variety of educational purposes. The underpinning idea of concept mapping is that meaningful learning can be realized by means of graphic representation, where the already-acquired knowledge pieces are organized into an integrated system, and this system opens to new knowledge as its future components [1,10,25,26]. Novak and Gowin [1] refine three fundamentals of Ausubel's learning theory as the base of concept mapping:

- Hierarchical structure refers to knowledge as part of an inclusive and systematic framework.
- Progressive differentiation is defined as engendering new concepts and ideas as knowledge deepens.
- Integrative reconciliation elucidates interrelationships between concepts.

Concept mapping has been regarded as an effective teaching and learning tool in accounting education. Relevant studies fall into to three schemes: making instructor-prepared accounting concept maps, teaching accounting students how to make concept maps, and assessing effectiveness of concept maps. These research themes are shown in the Table 1.

In the first scheme, the researchers themselves act as the creators of concept maps and use these maps as educational tools for the purpose of enhancing students' understandings of academic content. In the second scheme, the researchers, as concept map educators, equip students with the knowledge of how to design concept maps. Consequently, students create their own concept maps from a particular topic of accounting, and they are expected to be able to use this useful tool in other fields of accounting and non-accounting disciplines. The third scheme targets assessing the effectiveness of concept maps in contributing to the improvement of education quality; this is normally done by comparing students with experience using concept maps in the curriculum with those who have no such experience.

**Table 1.** Research schemes of concept mapping in accounting education.

| Research Schemes | Articles |
|---|---|
| Making instructor-prepared concept maps to enhance learning | Simon (2007) designs a curriculum concept map for Accounting Theory [10]; Handy and Polimeni (2017) propose six concept maps that correspond to six topics of the Introductory Managerial Accounting curriculum [27]; Mass and Leauby (2014) introduce a concept map for the topic of financial reporting standards in the financial accounting curriculum [28]. |
| Teaching accounting students to produce concept maps | After informing students about the steps of concept mapping, Shimerda (2007) presents student-prepared concept maps for the accounting equation topic, then indicates usefulness of concept mapping in accounting education [29]. Simon (2007) shows some instructor-prepared maps as the guidance for accounting students to create their own concept maps [25]. |
| Assessing effectiveness of concept maps in accounting | Mass and Leauby (2005) compare two groups of students undertaking two accounting topics of income statement and cash flow statement. Its finding claims that the group using concept maps had a better understating of the two topics than the other group taught under the traditional instruction method [30]. Leauby et al. (2010) appraise concept maps as a valuable learning tool as a result from the survey response from students who used the concept map method in the introductory accounting courses [11]. |

This paper appertains to the first research scheme in an attempt to redesign a concept map to improve the Accounting Theory curricula. Given that current literature has highly appraised the student-generated concept mapping activities and teacher-student interactive mapping process for their contributions to an active learning environment, teacher-prepared or expert-made concept maps maintain an essential role in teaching practices [5,9,31]. Schewendimann emphasizes that teacher-made concepts "could support integrative understanding" and that, compared to student-generated maps, teacher-made maps seemingly "have an equally positive effect on improving students' achievement" [5] (p. 82). In particular, students suffering weak verbal ability or insufficient prior knowledge benefit greatly from expert-made or instructor-prepared concept maps by recalling more central ideas from these maps than from texts [5].

The new curriculum concept map developed in this paper provides an alternative to the old version of mapping (Simon 2007), which draws on a narrow financial accounting perspective and makes weak presentations of the connections between concepts [10]. Simon's map is shown in Figure 1.

Simon's work stands as the singular concept mapping attempt for the Accounting Theory curriculum in previous literature, but it suffers from some serious limitations [10]. First, drawing on a financial orientation, this map has lost its comprehensiveness in depicting up-to-date curriculum knowledge system, where the accounting academy has highlighted non-financial development and theorized accountability as the accounting base [12–15].

Second, Simon's map fails to capture the intricate interconnections among accounting theories, concepts, and models. This problem places this map at a disadvantage in teaching some curriculum themes that draw on these interrelationships, such as accounting evolution and sector-specificity of accounting practices (that is, practices of the same accounting principles vary significantly across different accounting sectors).

Third, the linking phrases such as "can be" and "e.g." substantively used in this map are too simple, superficial, and unprofessional to effectively characterize the relationships between concepts. Canas, Novak, and Reiska comment that "excellent maps are explanatory, not descriptive"—that is, the quality of maps depends on their "depth of explanation" (p. 15) [8]. Therefore, the real problem for concept mappers is not an ability to identify concepts to be linked, but the ability to determine propositions to provide a clear explanation, including the professional narratives labeled on these links [8]. Kinchin highlights that concept mapping expertise is evidenced in "technical terms to apply in linking phrases to increase the explanatory power of the map" (p. 291) [32]. Simon's map tends to be of underdeveloped maps, which, as Kinchin critiques, "have used simple linking phrases to join the concepts together" (p. 298) [32].

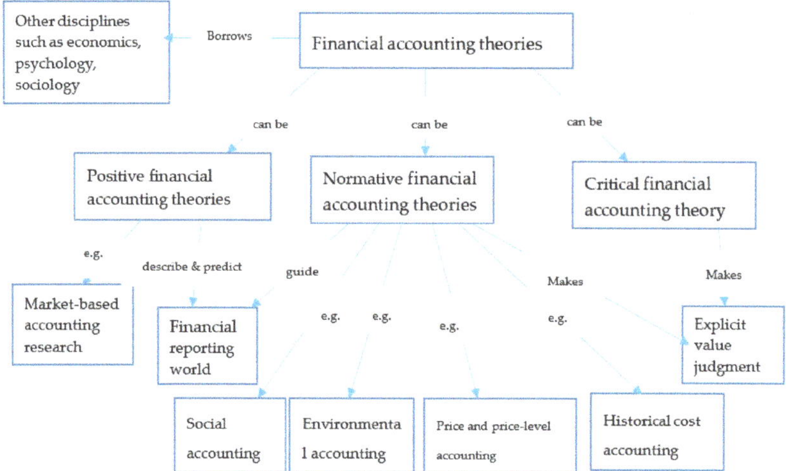


**Figure 1.** The old version of the curriculum concept map for (financial) Accounting Theory, sourced from Simon (2007) [10] with trivial modifications.

All the aforementioned problems are tackled in this paper by means of a new curriculum concept map using a tree model. How this concept tree is designed and how it differs from the previous mapping work are subjects further discussed in the following sections.

## 3. Design of the Concept Tree for the Curriculum of Accounting Theory

### *3.1. The Tree Concept Map and Accounting Theory Knowledge*

The tree form of concept maps, or the concept tree model, has been recognized in literature as one of the basic structures of concept maps [17,18]. Normally, concept trees are suitable for a knowledge system where concepts can be sorted into distinctly comparative classes and sub-classes. Recent development of the accountability perspective provides the opportunity to use a concept tree to (re)model the knowledge system of accounting theories. First, theorizing accountability as the base of accountings has become a generally accepted perspective and a robust thread throughout various accounting practices [13,15,33]. An accountability structure involves two interrelated social actors, called "accountor" and "accountee", where the accountor has an obligation to give an account that justifies her/his performance to the accountee [34–37]. As Shearer notes, "accounting practices enact accountability" (p. 569) [15], and accountability pervades the entire accounting phenomenon. To illustrate this, the ultimate purpose of traditional financial accounting is to enable a governing body of the reporting entity to discharge accountability to shareholders [37]. Comparatively, the objective of sustainability accounting is for the entity to discharge accountability to stakeholders [33,38–41].

Second, the diversification of accounting contexts/sectors is justified from an analytic perspective of accountability. Table 2 presents the accountability relationships that regulate practices in the three main accounting contexts/sectors, including corporate financial accounting, public-sector accounting, and sustainability accounting.

**Table 2.** Accountability and accounting contexts/sectors.

| Accounting Contexts/Sectors | Accountor | Accountee | Account |
|---|---|---|---|
| Corporate financial accounting | Corporation /business entity | Shareholders | General-purpose financial reports |
| Sustainability accounting | Corporation /business entity | Stakeholders | Formal and informal environmental and social reports |
| Public sector accounting | Government | The public | The government fiscal reports and audit reports |

Note: this table integrates knowledge claims from [33,35,42].

### 3.2. The Sketch of the Concept Tree

In this redesign, the concept tree comprises the root, the trunk, the branches, the twigs, and the sub-twigs, all of which correspond to different concepts in different levels of the curriculum knowledge system.

Root: accountability denotes the root of the knowledge system, for it is the base of all accounting practices.

Trunk: the trunk is symbolized by the traditional (corporate) financial accounting as the mainstream accounting knowledge sphere. Financial accounting acts as the prototype for other accounting systems, including public-sector accounting and sustainability accounting. Both non-financial accounting systems entail a stage of simulating the traditional financial accountancy. It has been a zealous belief that public sector accounting originates from, mirrors, and keeps in line with corporate financial accounting [43–45]. Likewise, sustainability accounting is developed from the track of the financial accounting system [12,46].

Branches: sustainability accounting and public-sector accounting are denoted as two branches diversified from the trunk, which stands for the mainstream (corporate financial) accounting. Their emergence and development are attributed to critical stances against the conventional accountancy, and standards and practical models associated with the two accounting systems are comparable to but significantly different from the financial accounting venue [12,44]. For example, sustainability accounting standards are inhered with non-financial and not-for-profit elements in establishing the accounting scopes, principles, and procedures [39,40]. Barton compares public-sector accounting and corporate financial accounting, attributing the considerable variances between the two systems to "differences in the roles and operating environment of governments and business" [44].

Twigs and sub-twigs: twigs signify academic schools and streams within an accounting context/sector (branch). Associated with the schools or streams are sector-specific standards, principles, and theories, which are denoted as sub-twigs. For example, the stream of capital market research is a twig of the financial accounting branch, including sub-twigs such as the efficient market hypothesis. There are lower levels of sub-twigs, such as the practical models to operationalize the standards and the principles. For example, the historical cost alternative (sub-twig) is operationalized into lower-level sub-twig models of economic value, net realized value, replacement cost, and deprival value.

### 3.3. Bifurcations

Bifurcations address the idea of "threshold concept", which refers to "a transformed way of understanding, or interpreting, or viewing something" (p. 1) [45]. A bifurcation symbolizes a threshold from which a new accounting branch/twig emerges from the conventional accounting venue. It marks the conflicting assumptions between two accounting branches or twigs that can be alternative to each other. The signposts labeled in bifurcations are used to express the rationales that underpin such inconsistencies and conflicts between concepts or theories. For example, the bifurcation between the financial accounting trunk and the sustainability accounting branch is labeled with the signpost, "financial versus non-financial", denoting the non-financial element that distinguishes sustainability accounting from its traditional counterpart. Another example is the twig-to-twig bifurcation with a sign of "historical cost versus anti-historical cost" to distinguish between financial reporting and the alternative measurement method. It suggests financial reporting is underpinned by the historical cost assumption, whilst the alternative measurement methodologists adopt an opposing stance to this assumption.

### 3.4. Linking Phrases

Twig notes are used as linking phrases to explain relationships between a higher level concept and a lower level concept, e.g., between a branch and a twig, or between a twig and its sub-twig. These twig notes (linking phrases) are shown in the Table 3.

**Table 3.** Twig notes as linking phrases.

| Twig notes | Abbreviation | Examples |
|---|---|---|
| Consist(s) of | C | Accounting principles consist of materiality. |
| Be derived from | D | Capital market research is derived from market efficiency hypothesis. |
| Be in the form of | F | Creative accounting can be in the form of "window dressing". |
| Be to guide | G | Financial reporting standards are to guide "recognition" of financial items. |
| Be operationalized as | O | Alternative costs can be operationalized as deprival value. |
| Regulates | R | Accountabiltiy regulates the "corporation–stakeholder relationship" in the domain of sustainability accounting. |
| Be to study | S | Agency theory is to study information asymmetry. |
| Be underpinned by | U | "New school" (of sustainability accounting) is underpinned by stakeholder primacy. |

### 3.5. Colors and Their Implications

Colors are used to highlight "accounting changes" from bifurcations, which result in germination of a new accounting branch or a new twig within the branch. We render further implications of colors used in this accounting tree, as shown in the Table 4.

**Table 4.** Colors and implications in the accounting concept tree.

| Objects | Color | Implications |
|---|---|---|
| Bifurcations | *Deep blue* | Deep blue represents wisdom. It implies potential to generate new sophisticated theoretical perspectives or philosophies. |
| Accountability | *Brown* | Brown is the color of soil, referring to accountability as the root for the whole accounting system. |
| Financial accounting (trunk) | *Gold* | The gold color denotes the for-profit assumption of (traditional corporate) financial accounting. |
| Sustainability accounting (branch) | *Green + Gold* | The new school is colored with green, aiming at improving environmental and social performance.<br>The old school adopts the same color of gold as financial accounting. This is in accordance with the view that old school holds a "for-profit" (shareholder primacy) element, and it is part of financial accounting. |
| Public-sector accounting (branch) | *Sky blue* | Sky blue stands for harmony and trustworthiness, associated with the "for the public" assumption of public-sector accounting. |
| Alternative measurement (twig) | *Yellow* | Yellow refers to opportunity and awareness, highlighting alternative measurement methods as "alternatives" to the traditional historical costing in making managerial decisions. |
| Capital market research (sub-twig) | *Red* | Red implies good signs or prosperousness in capital markets. |
| Creative accounting (sub-twig) | *Grey* | Creative accounting is a "grey zone" in financial reporting, exploiting loopholes in financial reporting regulation to gain advantages. It is not illegal but deviates from the spirit of accounting. |

### 3.6. The Curriculum Concept Tree of Accounting Theory

Figure 2 is the product of the tree-like concept map for the curriculum of Accounting Theory. In the figure, accounting theories and concepts are modeled into a tree. Located at the root is accountability, which represents the foundation of accounting practices. The accountability root buttressed the financial accounting trunk (main branch) and the branches of public-sector accounting and sustainability accounting. This layout visualizes the theory that accountability underlies all accounting phenomena, and different accountor–accountee relationships regulate various sector-specific accounting practices [33,42,46–48]. The bifurcation signposts indicate the difference between accounting sectors (branches), academic schools and streams (twigs), and sector-specific standards, principles, and practical models (sub-twigs and lower-level sub-twigs).

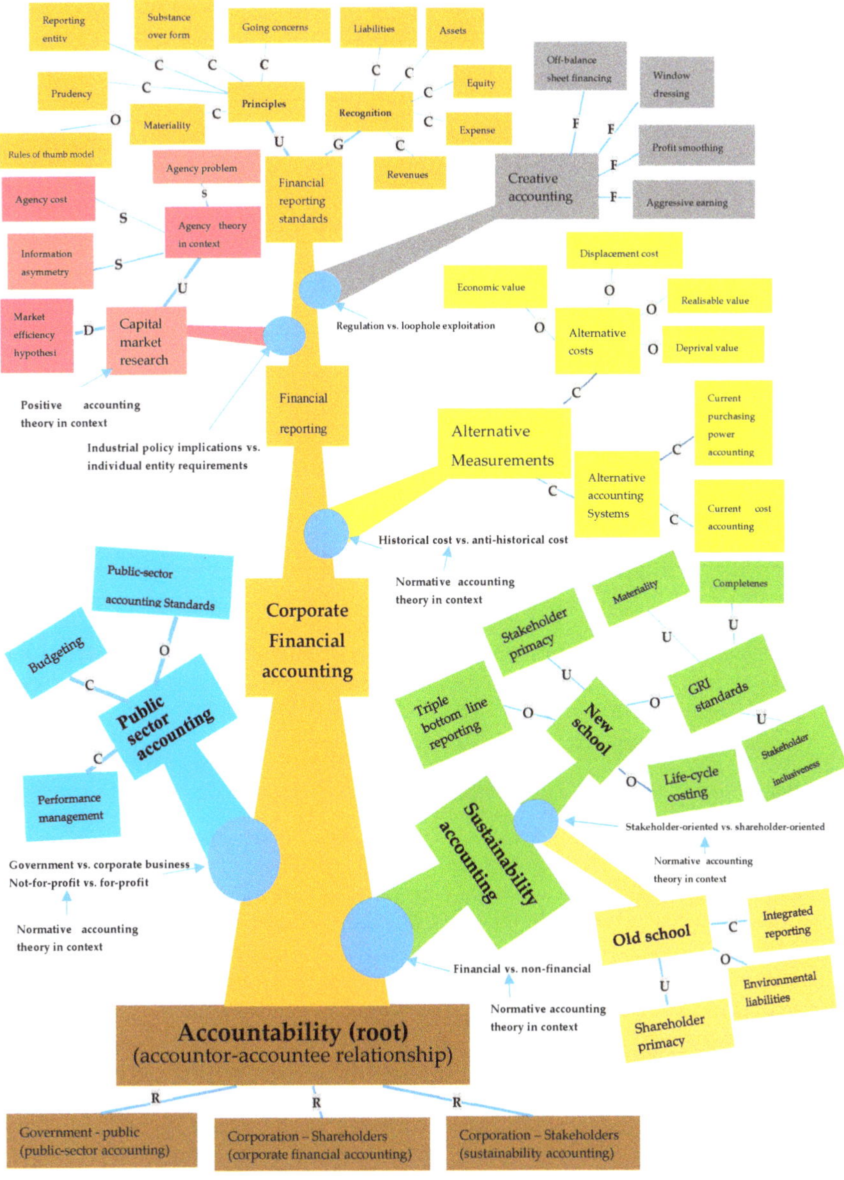


**Figure 2.** The concept tree of Accounting Theory.

### 3.7. A Comparison between the Two Curriculum Concept Maps

Table 5 outlines some main characteristics of this concept tree in contrast to Simon (2007)'s curriculum concept map of financial accounting theory.

**Table 5.** The main differences between Simon's map and the concept tree.

| Comparative Themes | The Old Concept Map (Simon 2007) [4] | The Concept Tree (Developed in this Paper) | Comments on the Redesign |
|---|---|---|---|
| Accountability | Not included. | Highlighted as the foundation of accounting practices. | Addressing the exceptionally important accountability perspective to accounting. |
| Public sector accounting | Not covered. | Public-sector accounting is diverged from traditional (corporate) financial accounting. | Addressing public-sector accounting that is not for profit and not applied in the business environment. |
| Environmental and social perspectives of accounting | Taken as part of financial accounting. | Sustainability (environmental and social) accounting is diversified from financial accounting and encompasses two streams. | The old school having a shareholder-primacy orientation is closely related to financial accounting, intimating the traditional view presented in Simon's map.<br>The new school holding non-financial and stakeholder-based elements has become mainstream in sustainability accounting. The new school is beyond the scope of Simon's map. |
| Comparison and contrast | It fails to recognize the alternative perspectives and lacks comparisons and contrasts between them. | Comparisons between concepts are used to indicate how accounting contexts/sectors and streams are diversified. | Addressing accounting evolution; justifying how new accounting systems and streams/schools were developed from the traditional venue. |
| Linking phrases | Focuses on only two simple terms, "can be" and "e.g.", to describe the relationships. | Uses multiple technical terms to indicate the relationships. | Elucidating the complex relationships between accounting concepts. |
| Accounting research methodology | Presented as individual streams. | Consolidated in accounting contexts/sectors. | Providing a concrete understanding on accounting research methodology. |

## 4. Educational Insights

Embedded with a research-centered element, the Accounting Theory curriculum is destined to assist students in comprehending the consolidation of accounting thoughts and discoveries, opening a door for them to probe into the nature of accounting phenomena. The concept tree plays an informative role to substantiate this essential aspect of the curriculum.

### 4.1. Feeling the Pulses of Accountancy Development

As an alternative to the old curriculum map [4], this concept tree embraces the advanced perspectives that have profoundly reframed the system of accounting knowledge. It enables students to update their theoretical understanding on accounting and then feel the pulses of accountancy development. In teaching practices, students are advised to compare the old and the new concept map and to clarify the major changes in accounting theories, including the general acceptance of the accountability-root view and the classification of accounting contexts/sectors that are regulated by different accountor–accountee relationships.

The elucidation is then prolonged according to this tree layout through introducing and commenting on the most recent theoretical debates associated with non-financial accountings, such as "business-like public sector accounting", "intelligent accountability", "theological or religious aspect of accountability", and "non-human stakeholders in sustainability accounting". These debates involve sharp critiques and radical ideas that have not been generally accepted or have remained exceptionally unpopular. However, they do represent cutting-edge understandings and brave attempts to promote advancement of accounting theories. Informed by the tree map, students not only acknowledge the new imperative accounting philosophies that have reformed the knowledge system, but also develop a precious spirit of critical thinking voiced from the infrequent endeavors in challenging the status quo.

### 4.2. Addressing Accounting Evolution

An advantage of this concept tree is its capacity to display and justify how accounting evolved and is evolving, which is attributed to the use of bifurcation signposts between different branches and between different twigs in a branch. To illustrate this, in contrast to the mainstream financial accounting (trunk) in pursuit of financial profits for individual business entities, public-sector accounting (branch) rests on a not-for-profit and public interest element, and sustainability accounting (branch) is developed to address non-financial and non-profit concerns. The bifurcation signposts between the trunk and the two new accounting branches characterize the base and the vein of how the two non-financial accounting systems emerged from critiques of the traditional accounting suppositions.

In a teaching setting, this concept tree is integrated in a story-telling method, where students can learn the accounting theories from a historical perspective. It was found that this innovative approach not only raised students' interests in studying theories themselves, but also triggered their enthusiasm in exploring the socio-economic backgrounds on which these theories were engendered. For example, presentation of the history of deprival value is complemented by the concept tree in the lecture. The bifurcation signpost between the twigs of financial reporting and alternative measurement informs that the deprival value contends an opposing stance against the historical cost assumption on which financial reporting standards are based. Associated with this bifurcation is a sequence of critical thinking questions regarding the invention of deprival value: (1) What were the problems of historical cost assumption when applied in practice? (2) How did the deprival value enable one to solve these problems? (3) How did the deprival value influence a generation of accounting scholars? (4) What lessons and experiences can we learn from this story regarding accounting evolution? These questions, ascribed to the historic story of deprival value, led students to probe into a deeper understanding of this accounting concept, motivating them to deliberate the nature and the practical implications of accounting evolution.

### 4.3. Understanding Accounting Research Methodology in Contexts

The distinction between the two research methods—normative accounting and positive accounting—is basic knowledge in this curriculum. Compared to the traditional way of paralleling the accounting research methodology to the practical accounting systems (as shown in Simon's map (2007) [10], this concept tree integrates the methodological understanding across and within accounting contexts. For instance, the concept tree illustrates the origin of positive accounting in the field (twig) of capital market research [49,50], and the normative accounting research is associated with the critique of existing theories and the exploration of alternative theories (refer to the bifurcation signposts). This is evidenced in the emergence of new branches or twigs in the concept tree. Sustainability accounting can be understood as one example of normative theory, for this new accounting system with a non-financial element is used to substitute for the traditional financial accounting routines in dealing with non-financial environmental and social issues [12]. The alternative measurements can be viewed as another example of normative theory, in which the historical cost assumption is refused due to its inability to reflect current value and future benefit.

### 4.4. Enhancing Understanding of Sector-Specific Accounting Practices

The context- or sector-specificity of accounting practices as a relatively new philosophic perspective has gained prominence in the accounting academy [42,51,52]. It argues that, provided an accounting principle is applicable to different accounting sectors/contexts (such as financial business accounting, governmental accounting, and sustainability accounting), practical models and operationalization methods of this principle vary significantly across these accounting sectors [12,42,44,53]. The concept tree reinforces the understandability of this important perspective, guiding students to observe, deliberate, and analyze the sector-dependent nature of accounting phenomena. For example, the GRI sustainability reporting standards and the financial reporting standards symbolize sub-twigs from different accounting branches in the concept tree. This graphic outlook enables an immediate grasp of the differences between the two reporting systems and provides a foundation for the later implementation of concrete cases, through which these differences are investigated in detail, from reporting principles to procedures and formats.

Another application exemplar is the research case of operationalizing the materiality principle in practices. The accounting academy has observed that public-sector accountants and auditors tend to apply a much stricter materiality level (significance of an issue) than their corporate financial peers [53,54]. That is, an issue that is considered insignificant from the view of corporate accountants and auditors may be regarded as significant for public-sector accounting practices. Theorists attribute this variance to the different scopes of accountability, where financial accounting practitioners are accountable to only a small group of shareholders, whilst public-sector accountants/auditors bear accountability to the whole society [53,55]. The concept tree enables a graphic reference to understand this phenomenon, where the materiality sub-twigs appear in the different branches of traditional financial accounting and public-sector accounting, and the two branches are regulated by different accountor–accountee relationships. This graphic reference drawn from the concept tree has been used to introduce the sector-specific materiality phenomenon and the relevant theoretical explanations, thereby enhancing a meaningful learning manner through which the theories are visualized [21,22].

## 5. Notes to the Curriculum Design

The concept tree performs a function of designing and steering the curriculum. The curriculum design experiences pertinent to its application are documented as follows.

### *5.1. Sequence of Lecture Topics*

Concept mapping is a useful vehicle for educators to "consider sequencing of topics" (p. 305) [11], a critical requirement for curricula due to the linear nature of instructional processes [1]. The concept tree endorses a sequence of curriculum topics. To begin, the concept tree is introduced in the initial lecture (Topic 1) to outline the overall curriculum knowledge system. Topic 1 also involves a brief explanation of the accountability theory and the accounting research methodology as they pertain to normative and positive theories. The other topics of the curriculum are sequenced and correspond to different parts of the tree, such as: financial reporting system (Topic 2, mainstream/trunk), alternative measurements to historical costs (Topic 3, twig), capital market research (Topic 4, twig), accountability theory (Topic 5, root), sustainability accounting (Topic 6, branch), public-sector accounting (Topic 7, branch), and advanced issues in accounting theory (Topic 8). Except for the introductory Topic 1, the sequencing starts from the financial reporting system (Topic 2) and its closely related Topics 3 and 4, and then to the non-financial accounting Topics 5, 6, and 7. The final topic, Topic 8, concerns progressive theses, antitheses, and debates in accounting studies, covering all previous topics.

The rationale inherent to this sequencing is in line with Ausubel's meaningful learning theory [21,22], which holds that an efficient learning process is produced when new knowledge can be related to knowledge already known. A prerequisite for the Accounting Theory (and Application) curriculum is the unit of Principles of Accounting, which conjoins with the topic of corporate financial reporting. As this topic contains knowledge that students are familiar with before they start the curriculum, it is set as the second topic immediately following the introductory topic. Accountability theory (Topic 5) is allocated before sustainability accounting (Topic 6). This arrangement is in accordance with an epistemic perspective that accountability theory is the ethical foundation for the development of sustainability accounting [13].

### 5.2. Cohesion of Topics

A problem pertains to the complex multifaceted content presented in the curriculum, where students often get lost in lectures due to difficulty relating individual knowledge pieces to the comprehensive picture [2]. To address this concern, the concept tree is displayed at the beginning of each lecture, highlighting the place of the current topic in the whole system of curriculum knowledge. During each lecture, students are inspired to use the bifurcation signposts to compare the current topic and the topics previously learned. For example, the bifurcation signpost, "industrial policy implications versus individual entity accounting requirements", indicates the difference of scope and purpose between the two topics/twigs—capital market research and financial reporting system. In the lecture of capital market research (Topic 3), this signpost provides hints for students to further develop their understanding by referring to Topic 2:

(1) how positive accounting researchers, whose work is to discover behavioral modes of companies in response to different accounting (Topic 3), play different roles from corporate accountants, who prepare financial reports according to the pre-set standards (Topic 2); and

(2) how to apply the agency theory in the capital market context (Topic 3) to prevent creative accounting problems associated with financial reporting standards (Topic 2).

### 5.3. A Meaningful Learning Mode of Exam Design

To our teaching experiences, exam questions prepared for this curriculum are easily designed on "a mode of rote learning" [21], which requires students to restate the definitions of accounting theories and concepts. The memory-test exam questions fail to assess whether students have really comprehended the theoretical knowledge and whether they are able to apply theories to practices. The concept tree can be useful to design consolidated case-based exam questions that integrate different accounting models, therefore attaining "the mode of meaningful learning" [21]. This application is shown in the following sample exam questions (Box 1).

**Box 1.** Memory-test (rote learning) questions prepared for the exam.

*Question 1: Define the normative accounting theory;*
*Question 2: Define the historical cost assumption;*
*Question 3: Discuss the advantages and the disadvantages of the historical cost method;*
*Question 4: Present the straight-line depreciation model to recognize a fixed asset;*
*Questions 5: Explain these values: market value, net realizable value, economic value.*

These exam questions simply require students to remember the textbook-based definitions and calculation formulas. The memory-test method to which the sample questions adhere is typical of rote learning and has been seriously critiqued in educational literature [1,21,22,56].

It is an attempt to redesign these questions in favor of a meaningful learning style. The first step is to establish a reference for these questions in the tree-shaped concept map, as shown in Table 6.

The bifurcation signpost, "historical cost versus anti-historical cost", indicates the fundamental differences between the two different twigs to which the financial reporting and the alternative costs belong, respectively. Furthermore, this signpost links to the adjacently noted "in-context normative theory", which was developed on the critique of the traditional historical cost. These connections form the condition under which it is possible to design a case used to integrate the separated knowledge pieces into a consolidated body for the exam, which is shown in Box 2.

Compared to exam questions in Box 1, which require students to replicate the definitions of individual concepts or formulas of the models, the exam redesign (Box 2) inspired by the concept tree synthesizes different knowledge pieces into a concrete, sophisticated, and practice-based case, which enables a more effective test of students' conceptual understandings and practical applications.

**Table 6.** Examined knowledge pieces in reference to the concept tree.

| Examined Knowledge Pieces | Reference to the Concept Tree |
|---|---|
| Financial reporting of a fixed asset | A lower-level sub-twig in the financial accounting standards twig |
| Calculation models of alternative values | Sub-twigs of alternative costs twig |
| Applications of the alternative values in decision-making | The twig of alternative costs |
| The historical cost assumption | The bifurcation signpost between twigs of financial accounting and alternative measurements |
| The positive and the normative accounting theory | In-context normative accounting theory associated to the twig of alternative measurement |

**Box 2.** A meaningful learning redesign of exam questions.

*On 1 January 20X5, a company purchased a machine at a cost of $2000. It has a useful life estimated to be 8 years, and this machine has a residual value of $400.*

*At 31 December 20X7, it was estimated that the machine could be sold for $4000. However, if the company could provide maintenance for this machine, it would be sold for $10,000. The maintenance fee was $2000. If the company continued to use the machine, management estimated that it would generate net cash inflows of $2000 each year during the remaining useful life.*

***Required:***

(1) *Calculate the amount of value for this asset shown in the financial report at the year-end 20X7.*

(2) *Calculate the market value, the net realizable value, and the economic value at the end of 20X7. At that time, the company had three options to deal with the machine: to sell the machine directly, to sell it after maintenance, and to keep it. What would be the best option from an economic view?*

(3) *Use this case to discuss the advantages and the disadvantages of the historical cost method to value a fixed asset. Suppose you try to write a research paper using this case in which you maintain a stance of advocating alternative costing and critiquing the traditional historical cost method. Is your research classified to positive or normative accounting theory?*

## 6. Conclusions

This paper is intended to extend the interpretational and the applicable capacity of concept mapping in accounting education, prompted by a knowledge gap that no suitable concept map has filled to interpret the current state of the Accounting Theory curriculum knowledge system, which has undergone rapid and significant development. The key molded by this paper to resolve this problem is a tree-form concept map, which systematically models the progressive curriculum knowledge into a tree-like structure, including the root, the trunk, the branches, the twigs, the sub-twigs, and the bifurcations.

The contribution of this study is trifold. First, it enriches the concept mapping methodology through an exemplary case of the scarcely documented "tree-shaped" mapping method and its application in an accounting education setting. This paper thus informs educators and concept mappers of the values and the characteristics of concept trees, conveying the potential to apply and generalize this particular mapping technique to other curriculums or courses.

Second, the concept tree itself denotes a creative theoretical model that concludes the accounting knowledge system, which has been significantly restructured and advanced by newly emerging theories. Compared to the previous limited mapping work [10], the concept tree assimilates the advanced and the prominent knowledge claims that accountability is theorized as both the base of accounting practices and as the rationale, according to which financial and non-financial accounting contexts/sectors are distinctly classified [12,44]. Furthermore, with a faculty of interpreting the complex interrelationships between concepts, this tree-shaped redisposition provides a reference to a deeper form of wisdom concerning the accounting evolution, the accounting research methodology, and the sector-specific accounting phenomena.

The curriculum development that the concept tree facilitates is the third contribution. The tree-shaped concept map sets visual guidelines for educators to sequence and cohere topics, thereby enabling students to concisely communicate the voluminous and complex information that encompasses many aspects of the curriculum [57]. In addition, this concept tree has proved to be a useful device in designing exam questions, by which the pedagogy is redirected from a memory-test mode toward a meaningful learning track, where students are required to make efforts to comprehend, integrate, and apply their theoretical knowledge into concrete practices.

**Funding:** This research received no external funding.

**Conflicts of Interest:** The author declares no conflict of interest.

## References

1. Novak, J.D.; Gowin, D.B. *Learning How to Learn*; Cambridge University Press: New York, NY, USA, 1984.
2. Canas, A.J. *A Summary of Literature Pertaining to the Use of Concept Mapping Techniques and Technologies for Education and Performance Support*; The Institute for Human and Machine Cognition: Pensacola, FL, USA, 2003.
3. Schwendimann, B. Concept mapping. In *Encyclopedia of Science Education*; Gunstone, R., Ed.; Springer: Amsterdam, The Netherlands, 2015.
4. Wise, A.M. Map It: How Concept Mapping Affects Understanding of Evolutionary Processes. Ph.D. Thesis, University of California, Davis, CA, USA, 2009.
5. Schwendimann, B.A. Concept maps as versatile tools to integrate complex ideas: From kindergarten to higher and professional education. *Knowl. Manag. E-Learn.* **2015**, *7*, 73–99.
6. Bridges, S.M.; Corbet, E.F.; Chan, L.K. Designing problem-based curricula: The role of concept mapping in scaffolding learning for the health sciences. *Knowl. Manag. E-Learn.* **2015**, *7*, 119–133.
7. Hay, D.B.; Proctor, M. Concept maps which visualise the artifice of teaching sequence: Cognition, linguistic and problem-based views on a common teaching problem. *Knowl. Manag. E-Learn.* **2015**, *7*, 36–55.
8. Cañas, A.J.; Novak, J.D.; Reiska, P. How good is my concept map? Am I a good Cmapper? *Knowl. Manag. E-Learn.* **2015**, *7*, 6–19.
9. Kinchin, I.M. Concept Mapping as a Learning Tool in Higher Education: A Critical Analysis of Recent Reviews. *J. Contin. High. Educ.* **2014**, *62*, 39–49. [CrossRef]
10. Simon, J. Concept mapping in a financial accounting theory course. *Acc. Educ.* **2007**, *16*, 273–308. [CrossRef]
11. Leauby, B.A.; Szabat, K.; Mass, J.D. Concept mapping—An empirical study in introductory financial accounting. *Acc. Educ.* **2010**, *19*, 279–300. [CrossRef]
12. Lamberton, G. Sustainability accounting—A brief history and a conceptual framework. *Acc. Forum* **2005**, *29*, 7–26. [CrossRef]
13. Roberts, J. No one is perfect: The limits of transparency and an ethic for 'intelligent' accountability. *Acc. Organ. Soc.* **2009**, *34*, 957–970. [CrossRef]
14. Schaltegger, S.; Burritt, R.L. Sustainability accounting for companies: catchphrase or decision support for business leaders? *J. World Bus.* **2010**, *45*, 375–384. [CrossRef]
15. Shearer, T. Ethics and accountability: from the for-itself to the for-the-other. *Acc. Organ. Soc.* **2002**, *27*, 541–573. [CrossRef]
16. Simon, J. Curriculum changes using concept maps. *Acc. Educ.* **2010**, *19*, 301–307. [CrossRef]
17. Vanides, J.; Yin, Y.; Tomita, M.; Ruiz-Primo, M.A. Using concept maps in the science classroom. *Sci. Soc.* **2005**, *28*, 27–31.
18. Solvie, P.A.; Sungur, E. Concept maps/graphs/trees/vines in education. Presented at the 5th WSEAS International Conference on E-ACTIVITIES, Venice, Italy, 20 November 2006.
19. Gilmour, S.J. Daz Sint Noch Ungelogenius Wort: A Literary Linguistic Commentary on Gurnemanz, Episode. In *Book III of Wolfram's Parzival*; Universitatsverlag Winter: Heidelberg, Germany, 2000.
20. Alexandre, F. Trees, waves and linkages: models of language diversification. In *The Routledge Handbook of Historical Linguistics*; Bowern, B., Evans, C., Eds.; Routledge: New York, NY, USA, 2014; pp. 161–189.
21. Ausubel, D. *The Psychology of Meaningful Verbal Learning*; Grune & Stratton: Oxford, UK, 1963.

22. Ausubel, D. *The Acquisition and Retention of Knowledge: A Cognitive View*; Springer: Amsterdam, The Netherlands, 2012.
23. Ausubel, D. The use of advance organizers in the learning and retention of meaningful verbal material. *J. Educ. Psychol.* **1960**, *5*, 267–272. [CrossRef]
24. Mayer, R.E. Can advance organizers influence meaningful learning? *Rev. Educ. Res.* **1979**, *49*, 371–383. [CrossRef]
25. Ausubel, D. *Educational Psychology: A Cognitive View*; Holt, Rinehart and Winston: New York, NY, USA, 1968.
26. Rovira, C. Theoretical foundation and literature review of the study of concept maps using eye tracking methodology. *Prof. Inf.* **2016**, *25*, 59–73. [CrossRef]
27. Handy, S.A.; Polimeni, R.S. Concept mapping—A graphical tool to enhance learning in an introductory cost or managerial accounting course. *J. Acad. Bus. Educ.* **2017**, *18*, 161–174.
28. Mass, J.D.; Leauby, B.A. Active learning and assessment: A student guide to using concept mapping in financial accounting. *Glob. Perspect. Acc. Educ.* **2014**, *11*, 41–63.
29. Shimerda, T.A. Concept mapping: A technique to aid meaningful learning in business and accounting education. *Indian J. Econ. Bus.* **2007**, *6*, 117–124.
30. Mass, J.D.; Leauby, B.A. Concept mapping—Exploring its value as a meaningful learning tool in accounting education. *Glob. Perspect. Acc. Educ.* **2005**, *2*, 75–98.
31. Cañas, A.J.; Novak, J.D. Concept mapping using cmap tools to enhance meaningful learning. In *Knowledge Cartography: Software Tools and Mapping Techniques*; Okada, A., Shum, S.B., Sherborne, T., Eds.; Springer-Verlag: London, UK, 2008; pp. 25–46.
32. Kinchin, I.M. Pedagogic frailty: A concept analysis. *Knowl. Manag. E-Learn.* **2017**, *9*, 295–310.
33. Gray, R.; Adams, C.; Owen, D. *Accountability, Social Responsibility and Sustainability: Accounting for Society and the Environment*; Pearson Education Limited: London, UK, 2014.
34. Goetz, A.; Jenkins, R. *Voice, Accountability and Human Development: The Emergence of a New Agenda*; United Nations Development Programme: New York, NY, USA, 2002.
35. Mashaw, J. Accountability and institutional design: Some thoughts on the grammar of governance. In *Public Accountability: Designs, Dilemmas and Experiences*; Dowdle, M.W., Ed.; Cambridge University Press: Cambridge, UK, 2006; pp. 115–156.
36. Schedler, A. Conceptualizing accountability. In *The Self-Restraining State: Power and Accountability in New Democracies*; Schedler, A., Diamond, L., Plattner, M., Eds.; Lynne Rienner Publishers: London, UK, 1999; pp. 13–28.
37. Bovens, M. Analysing and assessing accountability: A conceptual framework. *Eur. Law J.* **2007**, *13*, 447–468. [CrossRef]
38. Australia Auditing Standard Board. *Materiality and Audit Adjustments AUS 306*; Australian Accounting Research Foundation: Melbourne, Australia, 2001.
39. Global Reporting Initiative. *G4 Sustainability Reporting Guidelines (GRI G4)*; GRI: Amsterdam, The Netherlands, 2013.
40. AccountAbility. *AA1000 AccountAbility Principles Standard 2008 (AA1000APS 2008)*; AccountAbility: London, UK, 2008.
41. Murninghan, M. *Redefining Materiality II: Why it Matters, Who's Involved, and What it Means for Corporate Leaders and Boards*; AccountAbility: London, UK, 2013.
42. Zhou, Y. Materiality in Sustainability Accounting: A Critical Realist Perspective. Ph.D. Thesis, Southern Cross University, Lismore, Australia, 2017.
43. Anon. IFAC Public Sector Meets in New Zealand. *Acc. J.* **1993**, *72*, 34–37.
44. Barton, A. Professional accounting standards and the public sector—A mismatch. *Abacus* **2005**, *41*, 138–158. [CrossRef]
45. Meyer, J.; Land, R. *Threshold Concepts and Troublesome Knowledge: Linkages to Ways of Thinking and Practising within the Disciplines*; University of Edinburgh: Edinburgh, UK, 2003; pp. 412–424.
46. Christensen, M.; Newberry, S.; Potter, B. Enabling global accounting change: Epistemic communities and the creation of a 'more business-like' public sector. *Crit. Perspect. Acc.* **2019**, *58*, 53–76. [CrossRef]
47. Herath, G. Sustainable development and environmental accounting: the challenge to the economics and accounting profession. *Int. J. Soc. Econ.* **2005**, *32*, 1035–1050. [CrossRef]
48. Hardy, L.; Ballis, H. Accountability and giving accounts. *Acc. Audit. Acc. J.* **2013**, *26*, 539–566. [CrossRef]

49. Ball, R.; Brown, P. An empirical evaluation of accounting income numbers. *J. Acc. Res.* **1968**, *6*, 159–178. [CrossRef]
50. Watts, R.L.; Zimmerman, J.L. Towards a positive theory of the determination of accounting standards. *Acc. Rev.* **1978**, *53*, 112–134.
51. Modell, S. In defence of triangulation: A critical realist approach to mixed methods research in management accounting. *Manag. Acc. Res.* **2009**, *20*, 208–221. [CrossRef]
52. Modell, S. Critical realist accounting: In search of its emancipatory potential. *Crit. Pers. Acc.* **2017**, *42*, 20–35. [CrossRef]
53. Zhou, Y.; Zhou, G. Establishing judgement about materiality in government audits: Experience from Chinese local government auditors. *Int. J. Gov. Audit.* **2011**, *38*, 9–14.
54. Raman, K.; Van Daniker, R. Materiality in government auditing. *J. Acc.* **1994**, *177*, 71–76.
55. DeZoort, T.; Harrison, P.; Taylor, M. Accountability and auditors' materiality judgments: the effects of differential pressure strength on conservatism, variability, and effect. *Acc. Organ. Soc.* **2006**, *31*, 373–390. [CrossRef]
56. Novak, J.D. *Learning, Creating, and Using Knowledge: Concept Maps as Facilitative Tools in Schools and Corporations*; Lawrence Erlbaum Association: Mahwah, NJ, USA, 1998.
57. Edmondson, K.M. Concept maps as reflectors of conceptual understanding. *J. Res. Sci. Teach.* **1983**, *13*, 19–26.

*Article*

# Uncovering Types of Knowledge in Concept Maps

**Ian M. Kinchin [1,*], Aet Möllits [2] and Priit Reiska [3]**

1 Department of Higher Education, University of Surrey, Guildford GU2 7XH, UK
2 School of Educational Sciences, Tallinn University, Narva mnt 25, 10120 Tallinn, Estonia; aetm@tlu.ee
3 School of Natural Sciences and Health, Tallinn University, Narva mnt 25, 10120 Tallinn, Estonia; priit@tlu.ee
* Correspondence: i.kinchin@surrey.ac.uk

Received: 27 May 2019; Accepted: 7 June 2019; Published: 13 June 2019

**Abstract:** Concept maps have been shown to have a positive impact on the quality of student learning in a variety of disciplinary contexts and educational levels from primary school to university by helping students to connect ideas and develop a productive knowledge structure to support future learning. However, the evaluation of concept maps has always been a contentious issue. Some authors focus on the quantitative assessment of maps, while others prefer a more descriptive determination of map quality. To our knowledge, no previous consideration of concept maps has evaluated the different types of knowledge (e.g., procedural and conceptual) embedded within a concept map, or the ways in which they may interact. In this paper we consider maps using the lens provided by the Legitimation Code Theory (LCT) to analyze concept maps in terms of semantic gravity and semantic density. Weaving between these qualitatively, different knowledges are considered necessary to achieve professional knowledge or expert understanding. Exemplar maps are used as illustrations of the way in which students may navigate their learning towards expertise and how this is manifested in their concept maps. Implications for curriculum design and teaching evaluation are included.

**Keywords:** semantic density; semantic gravity; Legitimation Code Theory; expertise; theory-practice

---

## 1. Introduction

The primary focus of 21st century education is to support students to develop meaningful knowledge that can be applied to a range of evolving, real-world settings [1–3]. The world with all its complexity—including a rapid growth of information and knowledge, along with increased pressures on the educational system—creates a challenge to help students to develop the skills to navigate these complexities. Therefore, the key role of curricula at school and at university is to promote theoretical knowledge that underpins evolving practice, and to help students to navigate between theoretical and everyday knowledge and between different kinds of theoretical knowledge [3]. Additionally, learners in higher education have to be prepared with appropriate, authentic contextual knowledge to ensure graduate employability [4].

In any discipline, novices tend to have loosely organized knowledge, where concepts and strategies are not well linked, while experts have a highly organized and well-structured knowledge base that allows them to use information meaningfully to solve problems [5,6]. Rather than adopting a trial-and-error approach that is typical for a novice, we need experts that can use a principles-based approach to solve problems [7]. With this in mind, several researchers have demonstrated the benefits of concept mapping in teaching, learning and assessing scientific subjects. The use of concept maps has been shown repeatedly to be an effective tool for improving conceptual understanding [8–12], developing higher-order thinking skills [13], revealing misconceptions [14,15] and eliciting achievements and grades [16]. Therefore, we ask: what do concept maps reveal if we explore different types of knowledge (novice, theoretical, practical and professional) in students' concept maps?

The scoring of concept maps and the awarding of a single number to summarize map quality may give an indication of how much information a student has acquired during his/her study, but it does not provide any indication of the types of knowledge that have been acquired (e.g., conceptual or procedural) or the relationships that the student has identified between knowledge types. This recognition if different knowledges has been described as essential for developing the basic characteristic of the expert student [17] who needs to recognize the existence and complementary purposes of different knowledge structures. This has been overlooked in the research literature on concept mapping that has tended to foreground the development of conceptual knowledge to the exclusion of procedural knowledge. The focus on the development of a discrete single map structure has emphasized this bias in knowledge type, with procedural knowledge often being buried within a map of conceptual knowledge. Kinchin and Cabot have discussed how expertise requires the oscillation between linear structures of procedural knowledge and networked structures of underpinning conceptual knowledge, but they did not offer any framework to assess the relationship between the two or how this may evolve over time [18].

In this paper we explore how different types of knowledge are embedded within a concept map and interact to each other. Concept maps that represent learners' knowledge structures have been associated with meaningful learning theory [19] and the promotion of higher order thinking skills [13]. Here, we present a major shift in emphasis in concept map evaluation by considering the analysis of concept maps in relation to the semantics dimension of Legitimation Code Theory [20]. This not only provides a commentary on the student's progress, but also offers a critique of the curriculum experienced and the way in which it facilitates (or not) a student's development from novice to expert. Here, expertise is considered to be derived from the purposeful interaction of different knowledges (as described by [21]). We present examples of student maps that illustrate the way in which students may navigate the curriculum and argue that, in most cases, students do not reach the level of professional understanding.

The expert structure that represents professional knowledge is explicit in the integrated nature of theoretical knowledge and the way in which this underpins the procedural knowledge that constitutes the visible practice that defines a professional [22]. The derivation of chains of practice from theoretical knowledge is one of the hallmarks of expert knowledge [18]. However, we should not be surprised that this expertise is rarely exhibited by students, who grapple with their understanding of concepts before they are able to distinguish between conceptual and procedural knowledge, or that it is rarely depicted in concept maps that generally aim to combine procedural and conceptual knowledge within a single structure. The example of professional knowledge given in Figure 1 (of local anesthetics in dentistry) shows how knowledge that has a high semantic density and low semantic gravity, SD+SG- (such as physiology and pharmokinetics), determines the structure of the theoretical knowledge to the right, whilst the chain of practice to the left is composed of concepts such as instrument assembly and techniques, which exhibit lower semantic density and high semantic gravity (SD-SG+). In this paper, we explore the possibility of locating elements from the practical and the theoretical in students' emerging understanding of a discipline as an indicator of their current status on the journey through secondary and higher education towards professional knowledge.

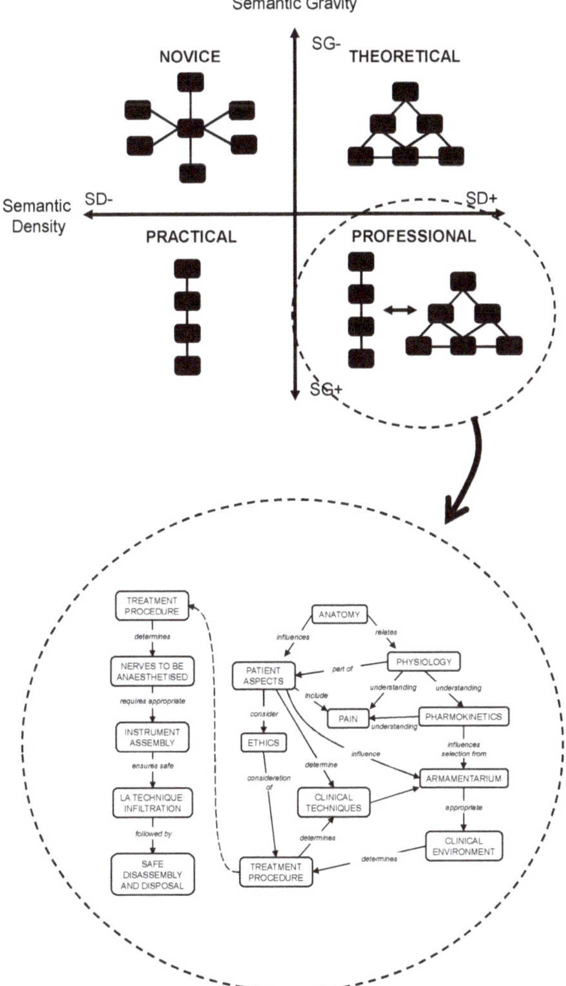


**Figure 1.** The semantic plane in which each quadrant has been populated by the archetypal map morphology (spoke, chain and network) that is likely to be found there, with (inset below) an example of a well-defined expert knowledge structure in which practice and theory are clearly delineated as complementary chain of practice and network of understanding [17,20,23].

## 2. Theory of Concept Maps

Concept maps have their roots in Ausubel's meaningful learning theory, and they emphasize the connections among concepts that represent individuals' knowledge structure [10,24,25]. There are three elements from Ausubel's theory that Novak and his research team found useful to develop in the concept mapping method:

(1) Construction of new meaning involves conceptual connections between new information and prior knowledge.
(2) Hierarchically organized cognitive structure where more general concepts are higher level in the hierarchy and less general are positioned under the more general concepts.

(3) Meaningful learning takes place when relationships between concepts are explicit and are better integrated with other concepts and propositions [10].

Concept maps are composed of concepts that are written in boxes and connected with arrows that are labeled to indicate the relationship between concepts [26]. The labeled connections between concepts are called links, and each 'concept-link-concept' forms a proposition that can be read as a stand-alone meaningful expression. Cross-links, which might sometimes be formed, show the relationships between two different areas of the map [27]. Concept mapping is a skill that encourages nonlinear thinking [28]. The construction process of concept mapping helps the learner to actively construct their knowledge and, as suggested by Hyerle [29], helps students to "think inside and outside the box". The important function of this graphic representation is to display the overall arrangement of concepts and the enhancement of metacognitive skills [7,12]. According to Salmon and Kelly, concept mappers with these skills are able to (1) define specific thinking process as recurring patterns; (2) support the transferring these patterns across disciplines; (3) guide the building of simple to complex mental models and (4) reflect how the frame of reference influences their meaning-making, thinking patterns and understanding [7].

Kinchin expresses the benefits of using concept maps by saying, "This is a tool that helps me not only to see how the students are putting ideas together (or not), but can also help the students to diagnose their own difficulties" [17]. Much school learning is achieved through rote learning, while using strategies like note-taking, rewriting the textbook pages, summarizing as bullet points and completing 'fill-the-gap' test that are not as productive as concept maps to develop well-organized knowledge. Thus, learners who are used to learning through rote learning find the higher level thinking that is required to construct a concept map challenging [13]. Concept mapping has also been proposed as a useful tool to support the learning of complex topics, where learners have fragmented understanding and might face difficulties integrating all components to form a meaningful overview [12]. The external scaffolding that the concept mapping process involves can be very helpful to support deep thinking and complex learning [7].

*Concept Maps—Hierarchy and Scoring*

Concept maps are unique for their graphical structures that exhibit how one concept is sub-ordinate to other concepts and how learners' understand the concepts [12,30]. A hierarchical concept map (also called a "Novakian concept map") is recognizable for its top-down fashion, where more general subordinate concepts are on top and more specific concepts are at the bottom. For instance, Novak and his colleagues claim, "A well-organized cognitive structure (which is necessary for meaningful learning) usually leads to graphically well-organized concept maps; in turn, building good concept maps helps to build a good knowledge structure" [31].

Several authors [11,30,32–34] associate the map hierarchy with the learning context. As stated earlier, the propositional structure is an essential part of concept mapping and shows learners' meaningful learning. However, not all 'concept-link-concept' triads form a meaningful proposition because they might miss the proper structure, have no logical meaning or constitute a large grammatical structure (e.g., sentence) that has no meaning independently within this bigger structure [35]. There are many authors, who consider different aspects of quality and complexity of concept map structure within their scoring rubrics.

The semantic scoring rubric of Miller & Cañas consists of six key criteria that are inherent for all concept maps [35]: (1) the presence of focus question and root concept, (2) the correct propositional structure—link reworking and overall map reorganization; (3) the presence on inaccurate propositions (misconceptions); (4) the presence of dynamic propositions that involves, movement, action, change of state or dependency relationships (e.g., roots absorb water, electric charge generates electric fields, etc.); (5) the number of quality cross-links that establish correct, suitable, and instructive relationships and (6) the presence of cycles in which the direction of the arrows allows traversing the entire closed path in a single direction. All of these six levels are also translated to the content-quality scale that is followed by the categories of unevaluated, very low, low, intermediate, high and very high.

Other studies have suggested that the structure of the concept map carries important information about the understanding and quality of learners' knowledge [11,12,30,36]. Many authors [7,30,37–40] emphasize the effectiveness of the qualitative scheme that differentiates three morphological types of concept map categories—spoke, chain and network [11]. Their model is based on map morphology that has following characteristics [12,41,42]:

**(1)** Spoke graphical structure—(a) concepts form only a single level and all subordinate concepts are in relation to the root concept; (b) subordinate concept are not connected to the neighboring subordinate concepts; (c) deleting concepts from the map (except deleting the root concept), does not impact the overall structure; d) the links that are built-in to the spoke structure are simple, do not create cross links and do not impact neighboring subordinate concepts.

**(2)** Chain graphical structure—(a) the root concept is linked to the subordinate concept and forms a sequence with the next concepts. There is no hierarchy, but concepts are listed in multiple levels in relation to the root concept; (b) subordinate concepts are connected only with the next following concept; (c) deleting concepts impacts only the subordinate concept lower down in the sequence; (d) the links are compound and therefore the meaning is readable only as a whole.

**(3)** Network graphical structure—(a) concepts are related to the root concept and form multiple levels defined as a "highly integrated and hierarchical network (of concepts) demonstrating a deep understanding of the topic" [11]; (b) removing or adding concepts does not impact the overall structure, as the cross links maintain the integrity of the map; (c) network is structured across different levels with interconnections, and indicates deep understanding and meaningful learning strategies.

Extreme versions of each of these morphological types are depicted within the quadrants of the semantic plane (Figure 1) to indicate the stereotypical structures that may be found to depict novice knowledge, theoretical knowledge and practical knowledge. However, each of these extremes is not 'fixed' and may evolve into another in response to student learning. For example, a spoke structure may develop into a chain or a network over a period of time as the student's understanding develops and is more systemized and complex in response to further learning [12]. Besides that, Kinchin discusses what is a "good" and "poor" map by comparing the exam results with the maps [12,43]. He concludes that "poor" maps are not always indicators of poor performance and "good" maps not always predictors of good performance. There is no one common determination whether a concept map is really good in terms of indicating the presence of a sophisticated understanding. In addition, Kinchin [17] claims, "bigger does not always mean better when evaluating concept maps."

Cañas [31] uses the idea of an "excellent map," and considers that both content and structure are important to determine the map quality. Cañas and colleagues [31] describe excellent concept maps as being concise and explanatory, exhibiting a high degree of clarity and presenting a clear message. In addition, excellent maps should also be well balanced, well-structured and demonstrate learners' understanding.

## 3. Materials and Methods

Exemplar student concept maps (here constructed by students during school science lessons) are translated into commentaries on the types of knowledge depicted by converting the linking phrases between concepts into descriptions of their semantic density and semantic gravity (see Figure 2). In terms of semantic gravity (SG), each proposition is considered in relation to the way in which the student has articulated the degree to which the knowledge is either tied to a particular context (SG+) or offers a more generalizable view (SG-) (see Table 1). We distinguish here between knowledge that is very context bound (SG++) and that which is less tightly bound (SG+) to offer a more nuanced description of the knowledge quality. Propositions are also evaluated according to the semantic density that is depicted (SD), where students may be using simple, everyday descriptions in their explanations (SD-) or may be offering much more technical summaries that exhibit considerable condensation of meaning (SD+). Again, the degree of condensation is considered by using SD+, SD++, SD- and SD– (see Table 1). In this way, each of the quadrants of the semantic plane itself has four sub-quadrants into which propositions may be plotted, giving up to 16 variants across the semantic plane. Once each proposition has been translated to indicate its semantic profile (SD±SG±), this is then plotted on the semantic plane (Figure 2) to indicate the semantic range depicted within the map. When using this method, researchers may need to establish the degree of inter-rater reliability to decide on ++ or + and on – or -. In this study we had three authors who were familiar with the content and agreed upon the level of density and gravity within each proposition.

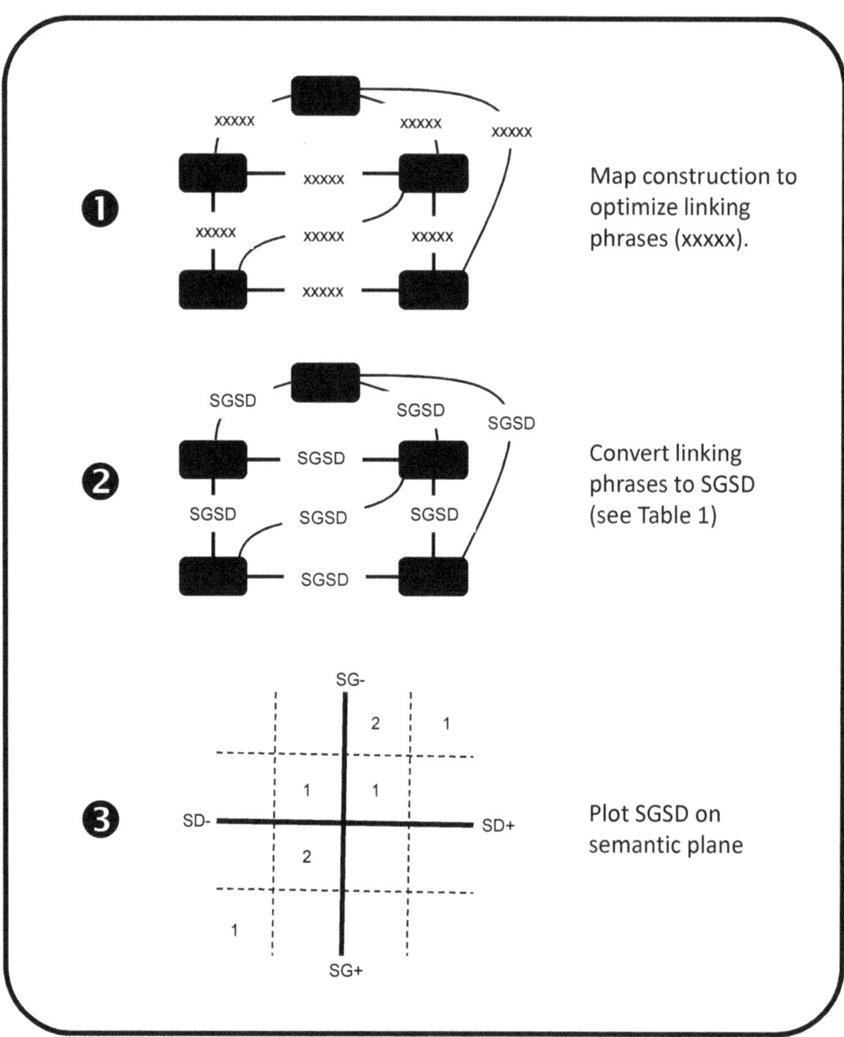


**Figure 2.** A three step process of map construction (1), translation (2) and plotting (3) on the semantic plane.

**Table 1.** Proposition analysis translation device. Modified from [44].

| | | |
|---|---|---|
| | **Novice Knowledge** | |
| SD-; SG- | **SD-**<br>- student needs to interpret only one concept to form a theoretically/scientifically correct proposition<br>- proposition does not need to be manipulated to the given context (the whole concept map) | **SG-**<br>- student uses concept from different sections of curriculum<br>- propositions create unified theory that is applicable to a broader context |
| SD-; SG– | **SD-**<br>- student needs to interpret only one concept to form a theoretically/scientifically correct proposition<br>- proposition does not need to be manipulated to the given context (the whole concept map) | **SG–**<br>- student uses abstract concepts (e.g., biology, chemistry, physics) and integrates them with general everyday knowledge that is applicable in a wide range of contexts<br>propositions might unify scientific principles by highlighting links between ideas |
| SD–; SG- | **SD–**<br>- student uses general everyday language and there is no theoretical knowledge needed to form a proposition<br>- forming a proposition does not need understanding or interpretation of scientific terminology (e.g., biology, chemistry, etc.) | **SG-**<br>- student uses concepts from different sections of curriculum<br>- propositions relate to ideas that are applicable to a broader context |
| SD–; SG– | **SD–**<br>- student uses general everyday language and there is no theoretical knowledge needed to form a proposition<br>- forming a proposition does not need understanding or interpretation of scientific terminology (e.g., biology, chemistry, etc.) | **SG–**<br>- student uses abstract concepts (e.g., biology, chemistry, physics) and integrates them with general everyday knowledge that is applicable in a wide range of contexts<br>- propositions might unify scientific principles by highlighting links between ideas |
| | **Theoretical Knowledge** | |
| SD+; SG- | **SD+**<br>- student uses specialized scientific concepts<br>- student needs to identify concepts before they can be interpreted to form a meaningful proposition | **SG-**<br>- student uses concepts from different sections of curriculum<br>- propositions relate to ideas that are applicable to a broader context |
| SD+; SG– | **SD+**<br>- student uses specialized scientific concepts<br>- student needs to identify concepts before they can be interpreted to form a meaningful proposition | **SG–**<br>- student uses abstract concepts (e.g., biology, chemistry, physics) and integrates them with general everyday knowledge that is applicable in a wide range of contexts<br>- propositions might unify scientific principles by highlighting links between ideas |
| SD++; SG- | **SD++**<br>- student needs to identify concepts (multiple steps required) to form a meaningful/scientifically correct proposition that interacts with the whole concept map | **SG-**<br>- student uses concepts from different sections of the curriculum<br>- propositions relate to ideas that are applicable to a broader context |
| SD++; SG– | **SD++**<br>- student needs to identify concepts (multiple steps required) to form a meaningful/scientifically correct proposition that interacts with the whole concept map | **SG–**<br>- student uses abstract concepts (e.g., biology, chemistry, physics) and integrates them with general everyday knowledge that is applicable in a wide range of contexts<br>- propositions might unify scientific principles by highlighting links between ideas |
| | **Practical Knowledge** | |
| SD-; SG+ | **SD-**<br>- student needs to interpret only one concept to form a theoretically/scientifically correct proposition<br>- proposition does not need to be manipulated to the given context (the whole concept map) | **SG+**<br>- student uses scientific concepts that are embedded in practical contexts<br>- proposition might express an example that is used commonly in everyday life |
| SD-; SG++ | **SD-**<br>- student needs to interpret only one concept to form a theoretically/scientifically correct proposition<br>- proposition does not need to be manipulated to the given context (the whole concept map) | **SG++**<br>- student uses scientific concepts that only require a recall of the definition or rule<br>- proposition expresses the knowledge that is located in a specific section of a curriculum |
| SD–; SG+ | **SD–**<br>- student uses general everyday language and there is no theoretical knowledge needed to form a proposition<br>- forming a proposition does not need understanding or interpretation of scientific terminology (e.g., biology, chemistry, etc.) | **SG+**<br>- student uses scientific concepts that are embedded in practical contexts<br>- proposition might express an example that is used commonly in everyday life |
| SD–; SG++ | **SD–**<br>- student use general everyday language and there is no theoretical knowledge needed to form a proposition<br>- forming a proposition does not need understanding or interpretation of scientific terminology (e.g., biology chemistry, etc.) | **SG++**<br>- student uses scientific concepts that only require a recall of the definition or rule<br>- proposition expresses the knowledge that is located in a specific section of a curriculum |

**Table 1.** *Cont.*

| Professional Knowledge | | |
|---|---|---|
| SD+; SG+ | **SD+**<br>- student uses specialized scientific concepts<br>- student needs to identify concepts before they can be interpreted to form a meaningful proposition | **SG+**<br>- student uses scientific concepts that are embedded in practical contexts<br>- proposition might express an example that is used commonly in everyday life |
| SD+; SG++ | **SD+**<br>- student uses specialized scientific concepts<br>- student needs to identify concepts before they can be interpreted to form a meaningful proposition | **SG++**<br>- student uses scientific concepts that only require a recall of the definition or rule<br>- proposition expresses the knowledge that is located in a specific section of the curriculum |
| SD++; SG+ | **SD++**<br>- student needs to identify concepts (multiple steps required) to form a meaningful/scientifically correct proposition that interacts with the whole concept map | **SG+**<br>- student uses scientific concepts that are embedded in practical contexts<br>- proposition might express an example that is used commonly in everyday life |
| SD++; SG++ | **SD++**<br>- student needs to identify concepts (multiple steps required) to form a meaningful/scientifically correct proposition that interacts with the whole concept map | **SG++**<br>- student uses scientific concepts that only require recall of the definition or rule<br>- proposition expresses the knowledge that is located in a specific section of the curriculum |

## 4. Results

The maps considered here were constructed by students aged 16–17 years in an Estonian high school. This data collection was a part of the large-scale study (LoTeGym) that was undertaken from 2012–2014 [45]. The concept mapping instrument was linked with interdisciplinary scenarios from a cognitive test. The test instrument consisted of four interdisciplinary everyday life related scenarios, where each focused on one science subject (biology, chemistry, geography and physics). The aim of the test was to evaluate students' ability to give a scientific explanation, pose scientific questions, solve scientific problems and to make reasoned decisions. Students were given 30 different types of concepts (science processes, everyday social issues-relates, etc.) to map on the topic of 'Milk—is it always healthy?' Some of these concepts were representations of 'everyday' knowledge (i.e., the practical application of the theoretical concepts derived from biology, chemistry and physics). After a period of training to see exemplar maps and to gain some familiarity with the software, a cohort of 187 students were given 45 min to construct a concept map. The concept mapping was carried out using the computer program CmapTools. To ensure consistency of the data collection, the introductory training sessions before the concept mapping task was undertaken by the same researcher. All students were given an example of how to construct a concept map before the main maps were constructed. One supervisor was in the classroom to assist with possible technical problems and to ensure adherence to the structural grammar of Novakian concept maps [26]. Whilst it was noted from preliminary observation that most of the maps display a gross morphology indicative of novice understanding (a spoke structure), there was a large degree of variation in the ways in which the concepts were arranged and in the quality of the propositions used to link concepts. From this cohort, two exemplars are illustrated below as worked examples to showcase the method for map analysis.

Figure 3 shows the map produced by one student. A quick observation indicates this to be a spoke-type map [11], in which chains of propositions radiate out from the central concept, but little cross-linking is evident between the chains. Once the propositions are converted to indicate the degree of semantic density and semantic gravity, it can be seen that >1/3 of the propositions are categorized as SG-SD- (indicative of novice knowledge). The remaining propositions are divided almost equally between the theoretical and practical quadrants of the plane, but none are ascribed to the lower right hand quadrant (professional knowledge).

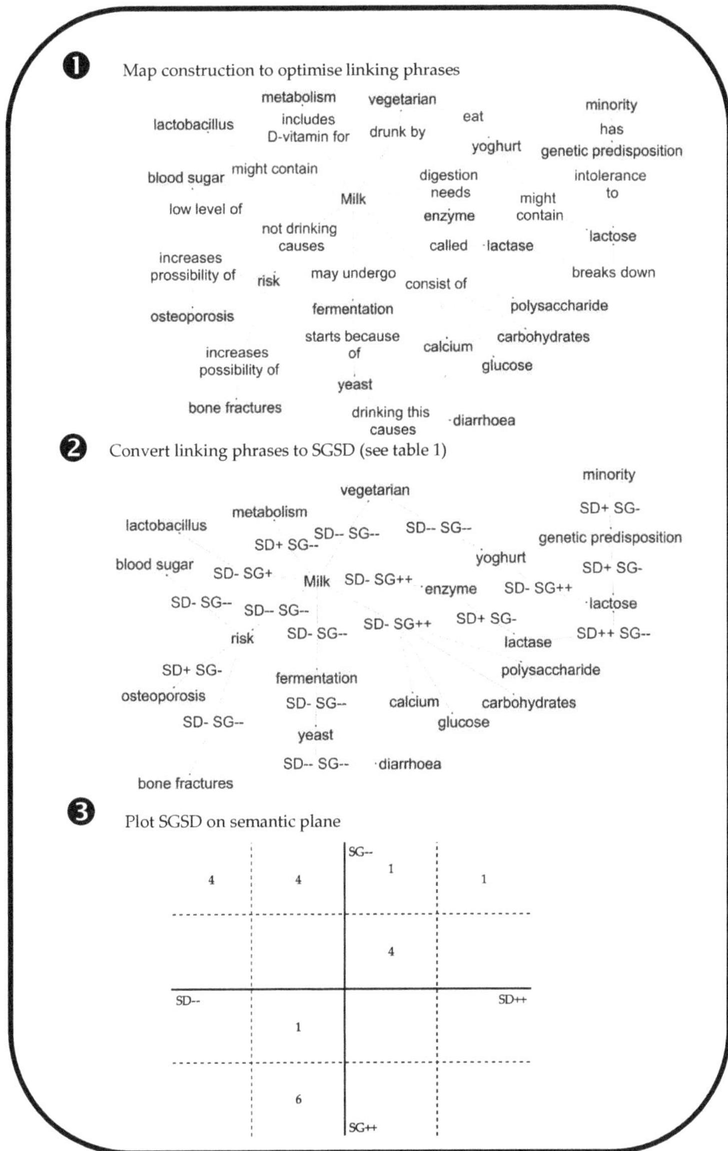


**Figure 3.** An example of a student map exhibiting a strong 'spoke' structure that suggests a novice understanding, which is emphasized by the presence of 8 propositions in the top left quadrant of the semantic plane.

The map in Figure 4 may also be designated as a novice map; however, there appears to be some development from the map in Figure 3, as the student here shows a greater attempt to show some cross-linking of concepts, moving from the spoke structure towards a more integrated network structure [11]:

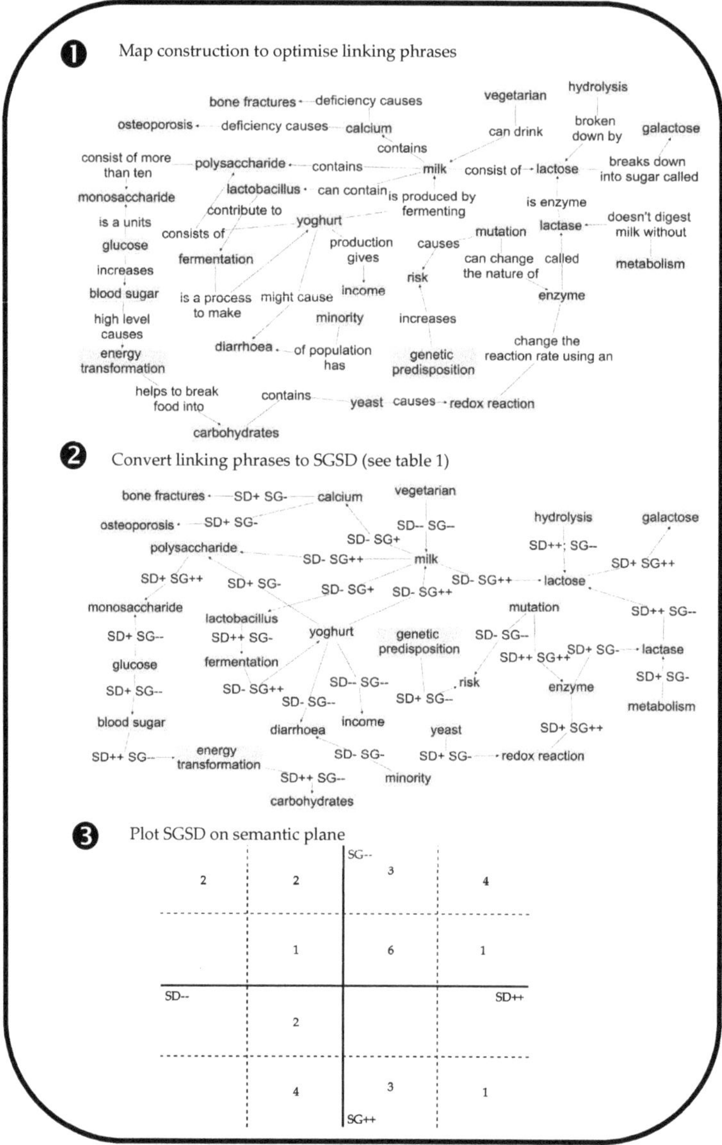


**Figure 4.** An example of a student map that suggests some emerging integration that builds on a novice structure, reinforced by the broad distribution of propositions across the semantic plane.

Whilst charting the position of the propositions across the semantic plane still indicates some novice knowledge (5 propositions), the majority of propositions represent theoretical (14 propositions) and practical knowledge (6 propositions), with some also being classified as professional knowledge. This suggests some semantic weaving on the part of the student.

## 5. Discussion

Disagreements within the research community about the most appropriate methods of analysis of concept maps have the potential to inhibit the widespread classroom use of the tool to support learning [43]. The benefits and drawbacks of traditional quantitative or qualitative approaches to map analysis are compounded by the fact that researchers have not previously discriminated between the types of knowledge that have been embedded within maps. The application of the semantics dimension of Legitimation Code Theory offers a new approach that is explicit in the need to consider understanding to be composed of qualitatively different knowledges that need to communicate with each other in the pursuit of expertise.

The consideration of the degree of semantic density and semantic gravity exhibited within map propositions offers a more nuanced consideration of map quality that is achieved by considering map morphology alone. However, it allows for the consideration of that which is 'yet-to-be-known' (rather than assessment of 'correctness') so that maps of contested values and beliefs can be assessed using the same approach as maps of agreed factual content [17]. The significance of the semantic profiles that students exhibit in their concept maps offers a window into some of the issues they experience on their educational journey—particularly as they move between school, university and professional life. For example, the differences in the structuring of knowledge that exist between a high school and a university biology curriculum (that has been observed by Kelly-Laubscher and Luckett [46]), suggest the existence of a possible mismatch between the semantic range that students are expected to navigate at university against that which they will have experienced in secondary school. This may cause problems for students' transition from school to university when their school education is assumed to have given them the necessary prerequisite knowledge to embark upon their undergraduate studies. Tracing the changes in the semantic profiles that students exhibit provides a visualization of the progress that students are making against desired outcomes, offering a way of monitoring student progression and curriculum effectiveness. However, we cannot assume homogeneity of the knowledge quality held by students as they enter university, even when they have covered the same content at school. The two examples shown here display differences in students' semantic profiles such that the student represented in Figure 4 appears to exhibit a greater semantic range within his knowledge structure of this content area, suggesting a better preparedness of undergraduate study. To confirm this, we need to explore a greater range of curriculum content with the students to see how key areas of the curriculum have been structured in the students' minds.

## 6. Conclusions

This new approach to concept map analysis raises a number of new opportunities and challenges for the research community:

By considering concept maps to be composed of different types of knowledge, it offers the possibility of asking a new set of research questions that might be addressed through concept mapping. Where powerful knowledge [47] is seen as the goal of professional education, then the semantic weaving between theory and practice is required to achieve expertise [21]. The assessment of this plurality of knowledges requires the mapping of semantic density and semantic gravity.

Beyond just assessing the 'correctness' of propositions within a map, the application of Legitimation Code Theory to concept mapping allows for the assessment of the ways in which the mapper is able to link theoretical knowledge with practical knowledge. This lifts the map above the assessment of factual recall and considers the higher order thinking skills that are required for students to achieve mastery of their discipline. This mastery has been shown to be dependent upon the learner's ability to oscillate between complementary knowledge structures consisting of chains of practice (exhibiting low semantic density and high semantic gravity), and underpinning networks of understanding (exhibiting high semantic density and low semantic gravity) [18,22]. The method of applying Legitimation Code Theory to concept mapping described in this paper provides a way to make the knowledges that underpin that expert practice explicit, so that they may be modeled for students. Further, this paper suggests that when assessing students' knowledge using concept maps, the use of a single map may be insufficient in order to obtain an authentic representation. As procedural and conceptual knowledge may be constructed differently and activated in different contexts, it may be better to encourage students to separate them structurally, whilst also recognizing the ways in which they interact in expert practice (as in Figure 1). This represents a significant methodological shift from many of the research papers that have previously explored learning using concept maps and that had assumed that complex knowledge may be captured in a single map structure.

**Author Contributions:** Conceptualization, I.M.K.; methodology, I.M.K. and A.M.; formal analysis, I.M.K. and A.M.; data curation, A.M.; writing—original draft preparation, I.M.K. and A.M.; writing—review and editing, I.M.K. and P.R.; supervision, P.R.

**Funding:** This research received no external funding.

**Conflicts of Interest:** Authors declare no conflict of interest.

## References


1. You, H.S.; Marshall, J.A.; Delgado, C. Assessing Students' Disciplinary and Interdisciplinary Understanding of Global Carbon Cycling. *J. Res. Sci. Teach.* **2018**, *55*, 377–398. [CrossRef]
2. Rimini, M.; Spiezia, V. Skills for a digital world: Background report 2016. *Knowl. Manag. E-Learn.* **2016**, *9*, 348–365.
3. Wheelahan, L. How Competency-Based Training Locks the Working Class out of Powerful Knowledge: A Modified Bernsteinian Analysis. *Br. J. Sociol. Educ.* **2007**, *28*, 637–651. [CrossRef]
4. Barnett, R. Knowing and Becoming in the Higher Education Curriculum. *Stud. High. Educ.* **2009**, *34*, 429–440. [CrossRef]
5. Ruiz-Primo, M.A. On the use of concept maps as an assessment tool in science: What we have learned so far. *Rev. Electrón. Investig. Educ.* **2000**, *2*, 29–52.
6. Bransford, J.D.; Brown, A.L.; Cocking, K.R. *How People Learn: Brain, Mind, Experience, and School*; National Academy Press: Washington, DC, USA, 2000; p. 38.
7. Salmon, D.; Kelly, M. *Using Concept Mapping to Foster Adaptive Expertise: Enhancing Teacher Metacognitive Learning to Improve Student Academic Performance*; Peter Lang: New York, NY, USA, 2015; p. 7.
8. Zimmerman, R.; Maker, C.J.; Gomez-Arizaga, M.P.; Pease, R. The Use of Concept Maps in Facilitating Problem Solving in Earth Science. *Gift. Educ. Int.* **2012**, *27*, 274–287. [CrossRef]
9. BouJaoude, S.; Attieh, M. The Effect of using concept maps as study tools on achievement in chemistry. *Eurasia J. Math. Sci. Technol. Educ.* **2008**, *4*, 233–246. [CrossRef]
10. Novak, J.D.; Cañas, A.J. *The Theory Underlying Concept Maps and How to Construct and Use Them*; Institute for Human and Machine Cognition: Pensacola, FL, USA, 2008; pp. 1–36. Available online: http://cmap.ihmc.us/docs/pdf/theoryunderlyingconceptmaps.pdf (accessed on 7 May 2019).
11. Kinchin, I.M.; Hay, D.B.; Adams, A. How a Qualitative Approach to Concept Map Analysis Can Be Used to Aid Learning by Illustrating Patterns of Conceptual Development. *Educ. Res.* **2000**, *42*, 43–57. [CrossRef]
12. Kinchin, I.M. Visualising Knowledge Structures in Biology: Discipline, Curriculum and Student Understanding. *J. Biol. Educ.* **2011**, *45*, 183–189. [CrossRef]
13. Cañas, A.J.; Reiska, P.; Möllits, A. Developing Higher-Order Thinking Skills with Concept Mapping: A Case of Pedagogic Frailty. *Knowl. Manag. E-Learn.* **2017**, *9*, 348–365.
14. Gouli, E.; Gogoulou, A.; Grigoriadou, M. A Coherent and Integrated Framework Using Concept Maps for Various Educational Assessment Functions. *J. Inf. Technol. Educ. Res.* **2017**, *2*, 215–240. [CrossRef]
15. Burrows, N.L.; Mooring, S.R. Using Concept Mapping to Uncover Students' Knowledge Structures of Chemical Bonding Concepts. *Chem. Educ. Res. Pract.* **2015**, *16*, 53–66. [CrossRef]
16. Karakuyu, Y. The Effect of Concept Mapping on Attitude and Achievement in a Physics Course. *Int. J. Phys. Sci.* **2010**, *5*, 724–737.
17. Kinchin, I.M. *Visualising Powerful Knowledge to Develop the Expert Student: A Knowledge Structures Perspective on Teaching and Learning at University*; Sense: Rotterdam, The Netherlands, 2016; pp. 15, 73.
18. Kinchin, I.M.; Cabot, L.B. Reconsidering the Dimensions of Expertise: From Linear Stages towards Dual Processing. *Lond. Rev. Educ.* **2010**, *8*, 153–166. [CrossRef]

19. Romero, C.; Cazorla, M.; Buzón, O. Meaningful Learning Using Concept Maps as a Learning Strategy. *J. Technol. Sci. Educ.* **2017**, *7*, 313–332. [CrossRef]
20. Maton, K. *Knowledge and Knowers: Towards a Realist Sociology of Education*; Routledge: London, UK, 2014.
21. Maton, K. Building Powerful Knowledge: The Significance of Semantic Waves. In *Knowledge and the Future of the Curriculum. Palgrave Studies in Excellence and Equity in Global Education*; Barrett, B., Rata, E., Eds.; Palgrave Macmillan: London, UK, 2014; pp. 181–197.
22. Kinchin, I.M. Accessing expert understanding: The value of visualising knowledge structures in professional education. In *Ensuring Quality in Professional Education*; Trimmer, K., Newman, T., Thorpe, D., Padro, F., Eds.; Palgrave McMillan: London, UK, 2019; Volume 2, pp. 71–89.
23. Clarke, F. Injecting expertise: Developing an expertise-based pedagogy for teaching local anaesthesia in dentistry. *High. Educ. Netw. J.* **2011**, *2*, 29–43.
24. Ausubel, D. *Educational Psychology: A Cognitive View*; Holt Rinehart: New York, NY, USA, 1968.
25. Novak, J.D.; Gowin, D.B. *Learning How to Learn*; Cambridge University Press: Cambridge, UK, 1984.
26. Novak, J.D. *Learning Creating and Using Knowledge: Concept Maps As Facilitative Tools in Schools and Corporations*, 2nd ed.; Routledge: London, UK, 2009; pp. 1–317. [CrossRef]
27. Novak, J.D.; Cañas, A.J. The Origins of the Concept Mapping Tool and the Continuing Evolution of the Tool. *Inf. Vis.* **2006**, *5*, 175–184. [CrossRef]
28. Crandall, B.; Klein, G.; Hoffman, R.R. *Working Minds: A Practitioner's Guide to Cognitive Task Analysis*; The MIT Press: Cambridge, MA, USA, 2006; p. 54.
29. Hyerle, D. *Visual Tools for Transforming Information into Knowledge*, 2nd ed.; Crowin Press: Thousand Oaks, CA, USA, 2009; p. 91.
30. Nousiainen, M.; Koponen, I. Concept Maps Representing Knowledge of Physics: Connecting Structure and Content in the Context of Electricity and Magnetism. *Nord. Stud. Sci. Educ.* **2010**, *6*, 155–172. [CrossRef]
31. Cañas, A.J.; Novak, J.D.; Reiska, P. How good is my concept map? Am I a good Cmapper? *Knowl. Manag. E-Learn.* **2015**, *7*, 6–19.
32. Vanides, J.; Yin, Y.; Tomita, M.; Ruiz-Primo, M.A. Using concept maps in the science classroom. *Sci. Scope* **2005**, *28*, 27–31.
33. Kinchin, I.M.; De-Leij, F.A.A.M.; Hay, D.B. The evolution of a collaborative concept mapping activity for undergraduate microbiology. *J. Furth. High. Educ.* **2005**, *29*, 1–14. [CrossRef]
34. Ingeç, S.K. Analysing concept maps as an assessment tool in teaching physics and comparison with the achievement Tests. *Int. J. Sci. Educ.* **2009**, *31*, 1897–1915. [CrossRef]
35. Miller, N.L.; Cañas, A.J. A Semantic Scoring Rubric for Concept Maps: Design and Reliability. In *Concept Maps Connecting Educators, Proceedings of the Third International Conference Concept Mapping, Tallinn, Estonia, 22–25 September 2008*; Tallinn University: Tallinn, Estonia, 2008; pp. 60–67.
36. Hay, D.; Kinchin, I. Using concept mapping to measure learning quality. *Educ. Train.* **2008**, *50*, 167–182. [CrossRef]
37. Yin, Y.; Vanides, J.; Ruiz-Primo, M.A.; Ayala, C.C.; Shavelson, R.J. Comparison of Two Concept-Mapping Techniques: Implications for Scoring, Interpretation, and Use. *J. Res. Sci. Teach.* **2005**, *42*, 166–184. [CrossRef]
38. Gerstner, S.; Bogner, F.X. Concept map structure, gender and teaching methods: An investigation of students' science learning. *Educ. Res.* **2009**, *51*, 425–438. [CrossRef]
39. Meagher, T. Looking inside a Student's Mind: Can an Analysis of Student Concept Maps Measure Changes in Environmental Literacy? *Electron. J. Sci. Educ.* **2009**, *13*, 85–112.
40. Subramaniam, K.; Harrell, P.E. An Analysis of Prospective Teachers' Knowledge for Constructing Concept Maps. *Educ. Res.* **2015**, *57*, 217–236. [CrossRef]
41. Hay, D.; Kinchin, I.; Lygo-Baker, S. Making Learning Visible: The Role of Concept Mapping in Higher Education. *Stud. High. Educ.* **2008**, *33*, 295–311. [CrossRef]
42. Kinchin, I.M.; Streatfield, D.; Hay, D.B. Using Concept Mapping to Enhance the Research Interview. *Int. J. Qual. Methods* **2010**, *9*, 52–68. [CrossRef]
43. Kinchin, I.M. Concept mapping as a learning tool in higher education: A critical analysis of recent reviews. *J. Contin. High. Educ.* **2014**, *62*, 39–49. [CrossRef]
44. Rootman-le Grange, I.; Blackie, M.A.L. Assessing Assessment: In Pursuit of Meaningful Learning. *Chem. Educ. Res. Pract.* **2018**, *19*, 484–490. [CrossRef]

45. Rannikmäe, M.; Soobard, R.; Reiska, P.; Rannikmäe, A.; Holbrook, J. Õpilaste loodusteadusliku kirjaoskuse tasemete muutus gümnaasiumiõpingute jooksul. *Eesti Haridusteaduste Ajakiri* **2017**, *5*, 59–98.
46. Kelly-Laubscher, R.F.; Luckett, K. Differences in curriculum structure between high school and university biology: The implications for epistemological access. *J. Biol. Educ.* **2016**, *50*, 425–441. [CrossRef]
47. Young, M.; Muller, J. On the powers of powerful knowledge. *Rev. Educ.* **2013**, *1*, 229–250. [CrossRef]

*Article*

# The Salutogenic Management of Pedagogic Frailty: A Case of Educational Theory Development Using Concept Mapping

**Ian M. Kinchin**

Department of Higher Education, University of Surrey, Guildford GU2 7XH, UK; i.kinchin@surrey.ac.uk

Received: 29 May 2019; Accepted: 21 June 2019; Published: 25 June 2019

**Abstract:** This paper explores the development of educational theory (pedagogic frailty) that has emerged through the application of concept maps to understand teachers' conceptions of their roles within the complex higher education environment. Within this conceptual paper, pedagogic frailty is reinterpreted using the lens offered by the concept of salutogenesis to place the model in a more positive frame that can offer greater utility for university managers. This development parallels changes in the consideration of mental health literacy (MHL) across university campuses and avoids misapplication of a deficit model to the professional enhancement of teaching quality. For a detailed explication of this wider perspective of pedagogic health literacy (PHL), the connections with related and supporting concepts need to be explained. These include 'assets', 'wellness' and a 'sense of coherence'. Links between these concepts are introduced here. This reframing of the model has used concept mapping to explore the relationship between two complex ideas—pedagogic frailty and salutogenesis. It emphasizes pedagogic health as a continuum operating between frailty and resilience. Brief implications for academic development are included.

**Keywords:** concept mapping; salutogenesis; pedagogic health; academic development; teaching

---

## 1. Introduction

The current literature on teaching in university is increasingly populated with references about stress and burnout among academics [1,2]. This should raise concerns about the physical and mental health of colleagues working within this system [3], and about the pedagogic health of the higher education system overall. Numerous stressors can be seen to act within the academy. For example, new academics report dissonance between expectations of their role and actual teaching experiences [4]. In addition, competing agendas within universities seem to be adding to the pressures of work [5], while political changes in the system appear to be at odds with the values that drew many academics into academia in the first place [6]:

> Academics are experiencing a growing sense of disconnection between their desires to develop students into engaged, disciplined and critical citizens and the activities that appear to count in the enterprise university. (p. 526)

This generates strong feelings among academics, such as those described so colourfully by Leitch [7] who talks about feeling as though she is "riding two horses at the same time, being propelled simultaneously in opposing directions" (p. 166). The negative consequences of too much stress within the university workforce have been summarised by Mtsweni [8] in his analysis of responses to stress among university administrators:

> the person may attempt to reduce the amount of information to be dealt with by opting for a simplified belief system which denies the true complexity of the issues involved. Typically

> this might entail a move towards polarised problem solving with a simplistic yes/no or right/wrong analysis. This diminished judgement can involve an increased personalisation of issues or a hostile egocentricity. In this case the sufferer can only see their limited viewpoint and begins to feel persecuted, interpreting neutral events as being directed at them. Lack of balance is completed by magnification and minimisation whereby trivial are given undue emphasis whilst key factors are played down or ignored. This unsupportable level of cognition eventually leads to fatigue and a state of under-alertness, characterised by forgetfulness, foggy thinking and disorganisation which may be wrongly attributed to a lack of motivation. (p. 20)

Within the context of teaching in higher education, these stress symptoms can be observed to be exhibited by colleagues and these can impede the innovative development of teaching and encourage the rise of 'play safe' classroom practices [9]. Such stress can lead colleagues to consider innovation in teaching practice in a binary manner as either 'good' or more typically 'bad' without considering the wider implications of change and possible benefits to students. The manifestation of hostile egocentricity referred to by Mtsweni can be observed through everyday comments such as, "It just won't work in our department—the management don't understand that we are a special case!" And finally, small changes to relatively minor procedural issues (e.g., which line-spacing should students use in their essays [10]) are often discussed extensively and with passion whilst the 'elephant in the room' is left for another time. The combined effect of these unproductive tensions and workplace stressors that cause 'foggy thinking and disorganisation' can result in an environment exhibiting pedagogic frailty, where elements of the teaching environment seem to be working in opposition to each other so that teachers retreat into a conservative status quo [9] that may be professionally unsatisfying and pedagogically unsound [11]. To address these problems, the model of pedagogic frailty is aligned to key aspects of salutogenesis, to make it more amenable to university managers as a developmental tool. This paper is aimed at those who influence or deliver teaching at university, including teachers, technicians, administrators and managers, as all these roles have an impact on the discourses of learning and on the student experience. Whilst the literature on teaching understandably tends to focus on teachers, it is evident that other roles have an impact on what goes on and how it is reported.

## 2. Pedagogic Frailty and Salutogenesis

The model of pedagogic frailty arose from a fortuitous confluence of personal and professional experiences with a theoretical exploration of university teaching [12]. This drew on three decades of work in teaching and academic development by the author that included several hundred structured teaching observations during which observed teachers often talked about the positive and negative factors influencing their teaching. The author also drew on professional examination of key factors influencing practice in the design of new academic programmes of teacher development [13]. The evolution of the model was also informed by personal encounters with clinical frailty [14] during which the overlap between the literature examining clinical experience and teaching experience became apparent, combined with the theoretical exploration of the visualisation of 'powerful knowledge' [15]. In combination, this gave rise to the conditions in which the model of pedagogic frailty could emerge (Figure 1).

By using the four key dimensions within the model (that have already been explored extensively in the literature [16,17]) to add structure to reflections on teaching practice, so that personal perspectives may be used as a basis for developmental dialogue (such as the example inserted below the model in Figure 1). Although the focus of research on pedagogic frailty considers the university system (i.e., the ways in which the various roles in the institution contribute to teaching), investigations need to start by uncovering the range of perceptions held by individuals within that system. Numerous case studies have revealed the variety of perspectives held by academics across the spectrum of academic disciplines from the arts to the sciences [17] and the ways in which tensions might develop resulting from conflicting perspectives. The model of pedagogic frailty focuses on four key areas:

- The nature of the discourse on teaching and learning and whether this concentrates on the mechanisms and procedures of teaching (timetabling, assessments, feedback, etc.) or on the underpinning pedagogy (teacher expectations, professional values, student learning approaches, etc.).
- The relationship between the pedagogy and the discipline and whether teaching offers an authentic insight to the discipline in terms of relating theory and practice.
- How the research within the department relates to the teaching in the department, and how these links are exploited in teaching strategies and made explicit in the programme.
- How the teaching is regulated and evaluated and what appreciation there is of the role of the individual academic in the decision-making processes of the institution.

In studies published so far, the model seems to resonate with university academics who readily relate to the idea of frailty and the notion that aspects of the professional environment can create tensions that impede the development of teaching practice [18,19]. However, the negative undertones of the term 'frailty' have been recognised in the clinical literature [20] and may be seen as problematic by university managers when considering the professional development of university academics as it suggests a deficit model. We may attempt to overcome this by adopting an assets-based approach to the consideration of the wider concept of pedagogic health as a continuum linking the extremes of frailty and resilience. The consideration of assets updates the clinical analogy from which pedagogic frailty emerged by offering a parallel to increased consideration of health assets [21,22] within a continuum of health as proposed by Antonovsky in his exploration of salutogenesis [23,24]. Salutogenesis is defined as the study of 'why' and 'how' people stay well [25]. Staying well is related to the ability of individuals to manage tension, that is, how they respond to stressors. The management of tension helps to maintain health. The pathological model is analogous to a deficit model of health, whilst salutogenesis pays more attention to the management of assets that contribute to wellness, and so can be seen to offer links with the ideas of pedagogic health as a continuum between the extremes of pedagogic frailty and pedagogic resilience.

How individuals manage tension and stress in their daily lives and stay well has been referred to as 'salutogenic functioning' [8]. Reframing in this way firstly requires us to discard the dichotomy of diseased/healthy in favour of Antonovsky's health-ease/dis-ease continuum (reframed here as frailty–resilience for the educational context). This is reflected by the application of concept mapping that enables us to visualise nuanced academic perceptions of their 'pedagogic health' [17] in which the diversity of perspectives is valued, and an inappropriate binary good–bad distinction is never made.

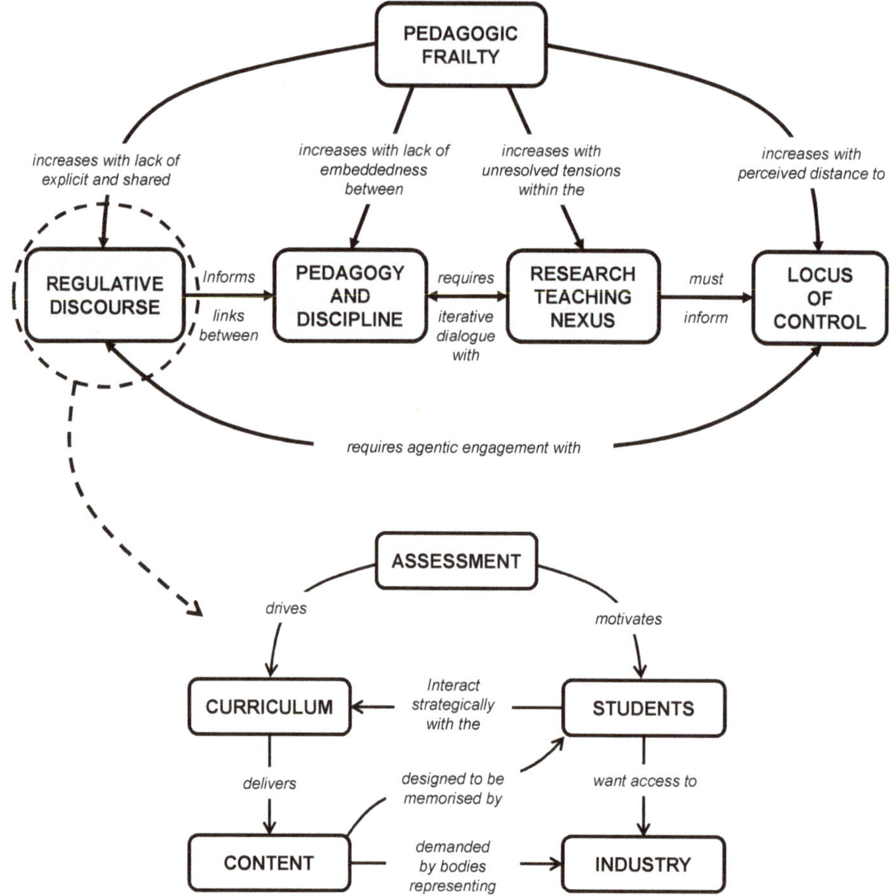


**Figure 1.** The overall pedagogic frailty model (above) with (inset below) one academic's view of the regulative discourse dimension (After Kinchin [15,26]).

Whilst most previous applications of salutogenesis in universities have been concerned with the physical or mental health of individuals working within a university [27], in this paper I have turned this around and am focusing on the health of the system (i.e., the university) where the individuals are working. However, these two perspectives are clearly related to each other and the distinction between a 'healthy academic' and a 'pedagogically healthy' university may be blurred across the numerous interactions between the individual and the institution. The consideration of salutogenesis as a frame for pedagogic health requires a parallel consideration of a number of other associated concepts (particularly assets, wellness and sense of coherence) that need to be part of the network of concepts that will help to generate a robust context to inform practice. Concept mapping offers a tool to allow the visualisation of these ideas and the ways they may be linked (Figure 2).

This visualization of the relationship between salutogenesis and pedagogic frailty represents the author's perspective of the main concepts involved and the relationships between them. The concept map was generated by reducing the problem to include only the main concepts involved and arranged to emphasize the relationships between them. The linking phrases have been constructed to offer the maximum explanatory power in the minimum amount of text in an attempt to produce what has been termed an 'excellent concept map' in the research literature [15]. This provides the reader with a map to complement the text as a way of reducing cognitive load and making the text more accessible when having to manage a set of unfamiliar terminology.

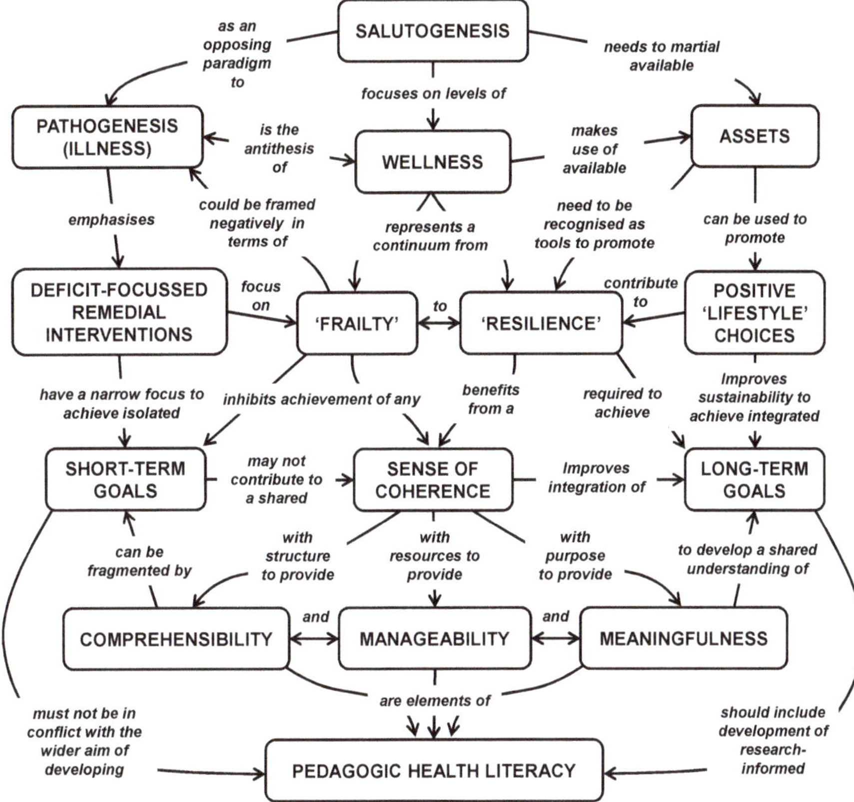


**Figure 2.** A concept map to illustrate the associated concepts that help to relate salutogenesis with pedagogic health literacy.

## 3. Assets

Health assets are starting to be a feature of the healthcare literature that provided the basis for the original frailty analogy. Rotegård et al. [21] define health assets as:

> the repertoire of potentials—internal and external strengths qualities in the individual's possession, both innate and acquired—that mobilise positive health behaviours and optimal health/wellness outcomes. (p. 514)

The assets that people bring to their professional teaching activities can be highlighted and consequently mobilised through reflective dialogue. Some assets may be part of the academic's disciplinary heritage and may be revealed though a process of conceptual exaptation, where familiar disciplinary concepts can be repurposed to support deeper and more personally relevant reflection on teaching (for disciplinary examples, see [28–30], where ideas such as care in nursing, contested concepts in politics and reactions in chemistry are repurposed to provide a language to consider teaching). This reflects the oft-quoted work by Ausubel [31], whose Assimilation Theory of Learning emphasises a constructivist epistemology in which the only place for further learning is provided by what the individual already knows as a basis for the construction of new knowledge. What academics know best is their disciplinary knowledge and ways of thinking. This echoes the view of frailty management offered by D'Avanzo et al. [32]:

> if we want frailty to be approached as a malleable and preventable condition, a bottom-up approach is needed [and] the tools through which frailty can be managed should come from [participants'] own context and resources. (p. 16)

Rather than providing a rather inert list of assets, a process of 'asset mapping' is suggested in the literature as a process that can help to emphasise the dynamic connections between assets as a way to increase their overall utility [21]. In the case of pedagogic frailty, this asset mapping probably needs to start from the individual perspectives held by academics (as in the case studies illustrated by Kinchin and Winstone [17]) which can then facilitate and structure the essential dialogue between members of an academic community [33,34] to start to map community assets. Distinguishing between individual, community or institutional assets may be helpful in operationalising the pedagogic health model and targeting resources to support the management of a developing sense of coherence [35].

## 4. Wellness

In parallel with the increased focus on health assets within the healthcare literature, there has been an increased focus on the concept of wellness as a way of moving towards health-promoting behaviours. Wellness has been defined by McMahon and Fleury [36] as:

> A purposeful process of individual growth, integration of experience, and meaningful connections with others, reflecting personally valued goals and strength, and resulting in being well and living values. (p. 48)

The idea of 'living values' clearly addresses the comment made by Manathunga in the introduction to this paper, whilst 'meaningful connections with others' is a necessary prerequisite for the mapping of shared assets (such as personal traits or disciplinary skills) mentioned above.

## 5. Sense of Coherence

By linking elements of innovative practice to the frailty model, we are able to support academics in their construction of a greater sense of coherence with regard to the fragmented and contradictory discourses of higher education. Developing a greater sense of coherence within academics of their teaching environments has always been one of the explicit intentions of the application of the pedagogic frailty model [17]. Within the salutogenic paradigm, Antonovsky [23] has defined the sense of coherence (SOC) in terms of its three subcomponents (comprehensibility, manageability and meaningfulness) as:

> a global orientation that expresses the extent to which one has a pervasive, enduring though dynamic feeling of confidence that:
>
> (1) the stimuli deriving from one's internal and external environments in the course of living are structured, predictable and explicable (comprehensibility);
>
> (2) the resources are available to one to meet the demands posed by these stimuli (manageability);
>
> (3) these demands are challenges, worthy of investment and engagement (meaningfulness) (p. 19)

The sense of comprehensibility is supported by consistent, structured information and is confounded by stimuli that are chaotic, random, accidental or inexplicable. Unfortunately, Brookfield [37] reports that some teachers describe their work to be 'bafflingly chaotic'—a situation that augurs badly for the development of a sense of coherence among university academics—presenting a challenge for university managers. A well-developed sense of coherence seems to be related to the ability of academics to cope with stress [38] and is likely to support the development of a positive approach to asset management and wellness.

## 6. Individuals in the System

One important difference between clinical frailty and pedagogic frailty (as previously made explicit in the literature) is that studies of clinical frailty have a focus on the wellbeing of the individual, and consider assessment and treatment of the individual, whereas pedagogic frailty has a broader focus on the system in which that individual operates. This means that any given configuration for an individual may initiate or promote pedagogic frailty in one environment, but promote resilience in another, more receptive, environment. This can be seen in particularly sharp contrast when academics move from one national context to another and find that assets that were valued at home are no longer recognised when they move abroad [39]. However, the structure of individual profiles might predict the potential for frailty, or in other words, certain scripts act as indicators of 'prefrailty'. In an extreme case, a hypothetical, stereotypic academic who is a new arrival at a university might state that "he doesn't care what his colleagues do, he will not adapt any aspect of his teaching to fit current fashions because he has been teaching for twenty years and has established an efficient routine that fits his lifestyle and allows his research to flourish". Such a person might be expected to have difficulties integrating into a new environment that might exhibit a more progressive attitude to innovative teaching approaches. More subtle issues might be predicted where academics map their perceptions of the dimensions of frailty and produce knowledge structures with morphologies that are undeveloped and do not provide sufficient structure to indicate critical reflection on practice. An additional difference between clinical and pedagogic frailty concerns age. Whereas clinical frailty is more prevalent in older patients, pedagogic frailty occurs at any stage of an academic's career and may be repeated as conditions change or as academics take on new roles [40,41].

Research suggests that frailty is a dynamic process that does not sit comfortably in the disease-centred paradigm that dominates medicine [42], and that there are opportunities along its pathway to transition out of, manage and/or prevent its adverse consequences. Considering clinical frailty, Gwyther et al. [43] write:

> Superficially, there appeared to be a dichotomy in beliefs about frailty management. On one hand, some policy-makers appeared to support a greater medicalisation of frailty, a need for frailty to be recognised as an authentic clinical issue by medical professionals and treated as such. On the other, there were views that frailty should be demedicalised and that frailty management should be conceived of as an adaptation to life stages and be embraced as a societal issue with ownership devolved to a wider societal network. (p. 4)

Again, there are direct analogies to be drawn from the comments above to the concept of pedagogic frailty. Rather than a medicalisation of pedagogic frailty, the modern educational world seeks to adopt greater managerialism and accountability to address any frailty in the system, so it may be 'treated'. This is exemplified by the classic "you said, we did" type of management response to student voice. The devolution of management offers a different strategy [44] that would decentralise ownership of the teaching environment that might facilitate frailty management as 'adaptation to professional life stages'.

## 7. Benefits of a Salutogenic Gaze towards Pedagogic Health

By adopting Antonovsky's salutogenic gaze [23,24] to reframe the recent literature on pedagogic frailty [16,17], we might consider the issues that act on teaching in terms of the broader concept of 'pedagogic health'. This requires a modification of the original model of pedagogic frailty (Figure 1) to emphasize the dynamic continuum between frailty and resilience (Figure 3). Introducing the concept of 'pedagogic health' and modifying the linking phrases within the model provides a subtle yet important development for a number of reasons, as it:

- Adopts a more affirmative language (pedagogic health literacy) that may be more appealing to senior managers, having a more positive subtext than frailty.

As an analogy, the increased recognition of mental health issues among both university staff and students has moved from a pathological model (dealing with problems after they have arisen) towards one advocating greater awareness of mental health literacy for all. One of the problems of dealing with student wellbeing within the current Higher Education environment is that 'students approach services when their mental wellbeing is already affecting their ability to cope' [45]. Rather than wait for problems to surface, it may be better to increase the mental health literacy (MHL) of everyone on campus as students with problems also have the potential to affect others including roommates, classmates and staff [3,46–48]. It is, therefore, an issue that affects us all, whatever our own state of mental health. Likewise, before waiting for academics to experience difficulties through frailty within their teaching, moving to the proactive promotion of greater pedagogic health literacy (PHL) across the campus is likely to have a more positive outcome for the institutional community.

• Avoids a potential misuse of the model through adoption of a simplistic harmful binary, the use of which to 'classify' staff would in itself be an indicator of prefrailty.

Within the managerial culture of the neoliberal university, there is pressure to find simplistic, instrumental measures that can be adopted for use as performance indicators [49]. The emerging body of work on pedagogic frailty has demonstrated an underpinning complexity to the teaching environment that cannot be adequately represented by a simple metric. This prevents the concept of pedagogic frailty (or pedagogic health) to be subverted for political means and to prevent the disconnections between expectations and practice described by Manathunga et al. [6].

• Indicates a continuum where no system is likely to exhibit 'total health' and so creates no arbitrary endpoint to prematurely terminate professional development.

The case studies of academics explored by Kinchin and Winstone [17] concentrate on academics who were already recognised as successful teachers. Therefore, each of them has the potential to contribute to pedagogic resilience within their institution. However, I note again here that individual success is not necessarily an indicator of resilience (rather than frailty) across the system, and that even the most successful teaching teams do not exhibit 'total health' (i.e., there is always something new to learn or a new skill to acquire). This depends on developing healthy, positive links between the individuals within a system (e.g., department) for that system to function well.

The learning and development of academics within this perspective do not have a predictable, linear trajectory with an easily defined or predicted endpoint. Rather, '[learning] is an entangled, nonlinear, iterative and recursive process, in which [academics] travel in irregular ways through the various landscapes of their experience (university, family, work, social life) and bring those landscapes into relation with each other' [50]. As such, it resembles the rhizomatic view of learning where knowledge is susceptible to constant modification as it responds to individual or social factors [51].

• The points listed above together help to make utilization of the model more 'management-friendly' and from which management activities are not removed.

It is assumed that senior managers may be reluctant to investigate frailty within the systems over which they preside, and of which they are an active part. The pathological model might be seen as a poisoned chalice. Therefore, by looking at pedagogic health, we have a perspective from which we hope senior managers would not feel the need to exclude themselves—something that would invalidate the whole enterprise.

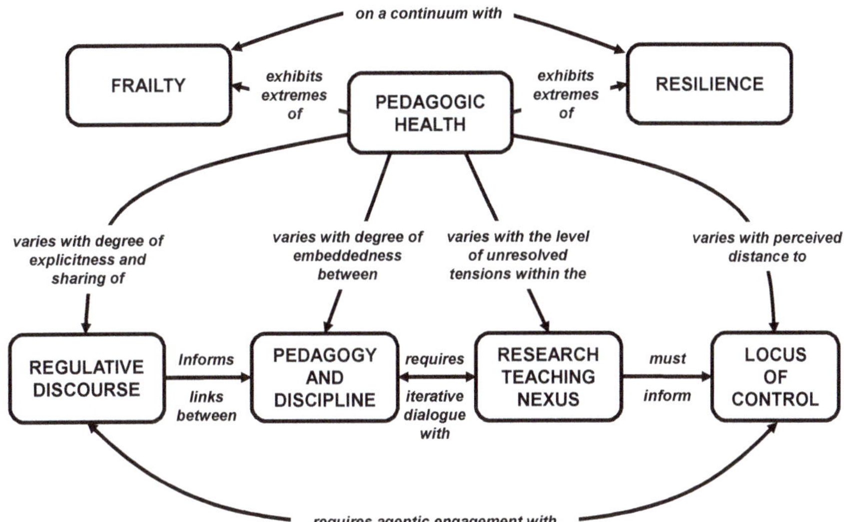


**Figure 3.** A revised model of pedagogic health to indicate the salutogenic continuum between extremes of frailty and resilience.

## 8. Conclusions

The development of a productive teaching environment is key to the success of a modern university. However, in the contemporary economic and political climate of higher education, there are many conflicting and competing discourses that need to be accommodated. Colleagues occupying different roles within the university structure (teacher, administrator, manager) will each have a different perspective on the university—what is important and what should be the institutional focus for the coming years. Inevitably, the development of teaching is a compromise between what we dream of being able to do and what is possible within the current constraints. In order for a university to move forward with a sense of coherence of purpose, the stakeholders within the institution need to have a shared vision of the key elements that make up the teaching environment. When this is achieved, we can talk about the pedagogic health of the institution in positive terms. Where this is not achieved, we might view the pedagogic health in less positive terms—tending towards frailty.

The adoption of appropriate language is an important issue when trying to initiate buy-in from academics (and their managers) to support interventions to enhance teaching quality. Use of the term 'frailty' in this context may invoke fatalistic connotations that were not the intention when originally applying the clinical analogy [26]. Buta et al. [52] have explored the language used in the literature surrounding clinical frailty and have identified some more positive expressions that can be used to better reflect the underlying philosophy of this work (for example, characterizing frailty as 'an opportunity for self-awareness and reflection') that sit better with the original pedagogic frailty model. This can be enhanced using a salutogenic gaze. In order for a salutogenic perspective on PHL to be operationalised within an institution, the network of understanding (Figure 2) needs to be related to productive chains of practice (*sensu* Kinchin and Cabot [53]), that will need to be context-specific. To this end, a modified model of pedagogic frailty is presented (Figure 3). In terms of practical application of this model, the diversity of perspectives revealed by concept mapping is important to help avoid a simplistic binary classification (frail–resilient) as that would preclude the development of individualised routes towards coping and adjustment [32]. Uncovering personal perspectives on the teaching environment [17] provides a shared lexicon to open up the potential for more collaborative discussions, helping to promote the development of a shared perspective underpinned by a set of common values [54]. This is likely to promote more productive dialogue about teaching, and the promotion of an environment that supports greater pedagogic health.

Within the author's university, moves towards a more salutogenic perspective of pedagogic health currently involve initiatives such as supporting a gradual transition from a 'responding-to-student-voice' mode of operation towards a greater degree of 'student–staff partnership' [55], as indicated in Figure 2 as a 'life-style choice' for the institution. Such a move is seen to have 'the potential to remedy neoliberal university models and performance self-regulation by offering a counternarrative to these dominant trends that imagine a different model of learning between students and staff' [56]. Analysis of the practicalities of developing a salutogenic gaze towards PHL across a university is not anticipated to be straightforward within institutions that have developed particular cultures and ways of doing things over many years. It is particularly important that the senior managers in an institution demonstrate a commitment to the counternarrative to ensure that the values that are espoused by the teachers working in partnership with the students are reflected in the language used by and actions demonstrated by the management [57].

The clinical literature continues to provide the basis for analogy in describing the variation in the way that frailty is experienced by individuals as they struggle to maintain their daily routines in the context of complicated transitions that create uncertainty [58]. Work on pedagogic frailty suggests a need to raise awareness among professionals (academics and university managers) of the malleability and preventability of frailty and the benefits of having an informed 'navigator' [43] or 'interpreter' [59]—which, in the case of pedagogic frailty, would be an academic developer [60] or learning developer [61]—to assist in steering a route through appropriate interventions. This provides a modified perspective in the role of the academic community as a whole, through a more distributed ownership of pedagogic health [44]. It also requires academic developers to be in a position to support the process of conceptual exaptation of disciplinary concepts for the enhancement of teaching. This may be achieved most effectively when academic development becomes a distributed activity [62], and those involved are actively engaged in research with their disciplinary peers [63]. This still requires effort on the part of all stakeholders to engage in professional development to ensure a positive direction of travel along the frailty—resilience continuum. As explained by D'Avanzo et al. [32], 'The resilience of people who succeed is achieved through continuous active development on numerous fronts' (p. 15).

In summary, it makes no sense to consider interrelated aspects of university teaching in isolation. Educational theory and analysis of practice needs to be considered as a connected whole. The salutogenic model of pedagogic health provides a lens to help achieve this goal, and concept mapping provides a tool that emphasises connectivity. Jonas [64] has referred to salutogenesis as an 'anchoring principle' to unify all dimensions of a healthcare system. I suggest here that salutogenesis offers great potential as a concept that can be repurposed to add clarity and a sense of coherence to the teaching endeavour. This added coherence is particularly important in the current context of political and economic uncertainty about the future of higher education and the need to care for our students [65]. To reinterpret comments by Becker et al. [66] by placing them into the discourse of teaching, 'A salutogenic gaze works prospectively by considering how to create, enhance, and improve resilience and provides a framework for researchers, and practitioners to help individuals, and organizations to move towards optimal pedagogic health' (p. 25). This potential seems worth exploring, and the application of concept mapping offers a nuanced methodology to do this.

**Funding:** This research received no external funding.

**Conflicts of Interest:** The author declares no conflict of interest.

## References

1. Darabi, M.; Macaskill, A.; Reidy, L. Stress among UK academics: Identifying who copes best. *J. Furth. High. Educ.* **2017**, *41*, 393–412. [CrossRef]

2. Mark, G.; Smith, A.P. A qualitative study of stress in university staff. *Adv. Soc. Sci. Res. J.* **2018**, *5*, 238–247. [CrossRef]
3. Gulliver, A.; Farrer, L.; Bennett, K.; Griffiths, K.M. University staff mental health literacy, stigma and their experience of student with mental health problems. *J. Furth. High. Educ.* **2017**, *43*, 434–442. [CrossRef]
4. McLean, N.; Price, L. Identity formation among novice academic teachers—A longitudinal study. *Stud. High. Educ.* **2017**, *44*, 1–14. [CrossRef]
5. Locke, P.; Maton, K. Serving two masters: How vocational educators experience marketisation reforms. *J. Voc. Educ. Train.* **2019**, *71*, 1–20. [CrossRef]
6. Manathunga, C.; Selkrig, M.; Sadler, K.; Keamy, K. Rendering the paradoxes and pleasures of academic life: Using images, poetry and drama to speak back to the measured university. *High. Educ. Res. Dev.* **2017**, *36*, 526–540. [CrossRef]
7. Leitch, R. On being transgressive in educational research! An autoethnography of borders. *Irish Educ. Stud.* **2018**, *37*, 159–174. [CrossRef]
8. Mtsweni, S.H. *Salutogenic Functioning Amongst University Administrative Staff*; University of South Africa: Pretoria, South Africa, 2010.
9. Bailey, G. Accountability and the rise of 'play safe' pedagogical practices. *Educ. Train.* **2014**, *56*, 663–674. [CrossRef]
10. Johansen, E.; Harding, T. 'So I forgot to use 1.5 line spacing! It doesn't make me a bad nurse!' The attitudes to and experiences of a group of Norwegian postgraduate nurses to academic writing. *Nurs. Educ. Pract.* **2013**, *13*, 366–370. [CrossRef]
11. Kinchin, I.M. Mapping the terrain of pedagogic frailty. In *Pedagogic Frailty and Resilience in the University*; Kinchin, I.M., Winstone, N.E., Eds.; Sense Publishers: Rotterdam, The Netherlands, 2017; pp. 1–16.
12. Kinchin, I.M. Pedagogic frailty: A concept analysis. *Knowl. Manag. E Learn.* **2017**, *9*, 295–310.
13. Kinchin, I.M.; Hosein, A.; Medland, E.; Lygo-Baker, S.; Warburton, S.; Gash, D.; Rees, R.; Loughlin, C.; Woods, R.; Price, S.; et al. Mapping the development of a new MA programme in higher education: Comparing private perceptions of a public endeavour. *J. Furth. High. Educ.* **2017**, *41*, 155–171. [CrossRef]
14. Kinchin, I.M.; Wilkinson, I. A single-case study of carer agency. *J. Nurs. Educ. Pract.* **2016**, *6*, 34–45. [CrossRef]
15. Kinchin, I.M. *Visualising Powerful Knowledge to Develop the Expert Student: A Knowledge Structures Perspective on Teaching and Learning at University*; Sense Publishers: Rotterdam, The Netherlands, 2016.
16. Kinchin, I.M.; Winstone, N.E. (Eds.) *Pedagogic Frailty and Resilience in the University*; Sense Publishers: Rotterdam, The Netherlands, 2017.
17. Kinchin, I.M.; Winstone, N.E. (Eds.) *Exploring Pedagogic Frailty and Resilience: Case Studies of Academic Narrative*; Brill: Leiden, The Netherlands, 2018.
18. Kinchin, I.M.; Alpay, E.; Curtis, K.; Franklin, J.; Rivers, C.; Winstone, N.E. Charting the elements of pedagogic frailty. *Educ. Res.* **2016**, *58*, 1–23. [CrossRef]
19. Kostromina, S.N.; Gnedykh, D.S.; Ruschack, E.A. Russian university teachers' ideas about pedagogic frailty. *Knowl. Manag. E Learn.* **2017**, *9*, 311–328.
20. Nicholson, C.; Gordon, A.L.; Tinker, A. Changing the way "we" view and talk about frailty. *Age Ageing* **2017**, *46*, 349–351. [CrossRef] [PubMed]
21. Rotegård, A.K.; Moore, S.M.; Fagermoen, M.S.; Ruland, C.M. Health assets: A concept analysis. *Int. J. Nurs. Stud.* **2010**, *47*, 513–525. [CrossRef]
22. Whiting, L.; Kendall, S.; Wills, W. An asset-based approach: An alternative health promotion strategy. *Commun. Pract.* **2012**, *85*, 25–28.
23. Antonovsky, A. *Unravelling the Mystery of Health: How People Manage Stress and Stay Well*; Jossey-Bass: San Francisco, CA, USA, 1987.
24. Antonovsky, A. The salutogenic model as a theory to guide health promotion. *Health Prom. Int.* **1996**, *11*, 11–18. [CrossRef]
25. Antonovsky, A. *Health, Stress and Coping*; Jossey-Bass: San Francisco, CA, USA, 1979.
26. Kinchin, I.M. Pedagogic frailty: An initial consideration of aetiology and prognosis. Paper Presented at the Annual Conference of the Society for Research into Higher Education (SRHE), Wales, UK, 9–11 December 2015. Available online: https://www.srhe.ac.uk/conference2015/abstracts/0026.pdf (accessed on 24 June 2019).

27. Dooris, M.; Doherty, S.; Orme, J. The application of salutogenesis in universities. In *The Handbook of Salutogenesis*; Mittelmark, M.B., Sagy, S., Eriksson, M., Bauer, G.F., Pelikan, J.M., Lindström, B., Espnes, G.A., Eds.; Springer: New York, NY, USA, 2017; pp. 237–245.
28. Derham, C. Nursing. In *Exploring Pedagogic Frailty and Resilience: Case Studies of Academic Narrative*; Kinchin, I.M., Winstone, N.E., Eds.; Brill: Leiden, The Netherlands, 2018; pp. 61–75.
29. Usherwood, S. Politics. In *Exploring Pedagogic Frailty and Resilience: Case Studies of Academic Narrative*; Kinchin, I.M., Winstone, N.E., Eds.; Brill: Leiden, The Netherlands, 2018; pp. 91–103.
30. Whelligan, D. Chemistry. In *Exploring Pedagogic Frailty and Resilience: Case Studies of Academic Narrative*; Kinchin, I.M., Winstone, N.E., Eds.; Brill: Leiden, The Netherlands, 2018; pp. 17–31.
31. Ausubel, D.P. *The Acquisition and Retention of Knowledge: A Cognitive View*; Kluwer Academic Publishers: Dordrecht, The Netherlands, 2000.
32. D'Avanzo, B.; Shaw, R.; Riva, S.; Apostolo, J.; Bobrowicz-Campos, E.; Kurpas, D.; Holland, C. Stakeholders' views and experiences of care and interventions for addressing frailty and pre-frailty: A meta-synthesis of qualitative evidence. *PLoS ONE* **2017**, *12*, e0180127.
33. Wiley, C.; Franklin, J. Framed autoethnography and pedagogic frailty: A comparative analysis of mediated concept maps. In *Pedagogic Frailty and Resilience in the University*; Kinchin, I.M., Winstone, N.E., Eds.; Sense Publishers: Rotterdam, The Netherlands, 2017; pp. 17–32.
34. Gkritzali, A.; Morgan, N.; Kinchin, I.M. Pedagogic frailty and conventional wisdom in tourism education. In Proceedings of the 7th International Critical Tourism Studies Conference, Palma de Mallorca, Spain, 25–29 June 2017.
35. Bonmatí-Tomás, A.; del Carmen Malagón-Aguilera, M.; Bosch-Farré, C.; Gelabert-Vilella, S.; Juvinyà-Canal, D.; Gil, M.D.M.G. Reducing health inequities affecting immigrant women: A qualitative study of their available assets. *Glob. Health* **2016**, *12*, 37. [CrossRef]
36. McMahon, S.; Fleury, J. Wellness in older adults: A concept analysis. *Nurs. Forum* **2012**, *47*, 39–51. [CrossRef] [PubMed]
37. Brookfield, S. Pedagogical peculiarities: An introduction. In *Pedagogical Peculiarities: Conversations at the Edge of University Teaching and Learning*; Medland, E., Watermeyer, R., Hosein, A., Kinchin, I.M., Lygo-Baker, S., Eds.; Sense Publishers: Rotterdam, The Netherlands, 2018; pp. 1–16.
38. Kinman, G. Work stressors, health and sense of coherence in UK academic employees. *Educ. Psychol.* **2008**, *28*, 823–835. [CrossRef]
39. Hosein, A.; Rao, N.; Yeh, C.S.-H.; Kinchin, I.M. (Eds.) *International Academics' Teaching Journeys: Personal Narratives of Transitions in Higher Education*; Bloomsbury Publishing: London, UK, 2018.
40. Lygo-Baker, S. The role of values in higher education: The fluctuations of pedagogic frailty. In *Pedagogic Frailty and Resilience in the University*; Kinchin, I.M., Winstone, N.E., Eds.; Sense Publishers: Rotterdam, The Netherlands, 2017; pp. 79–91.
41. Read, B.; Leathwood, C. Tomorrow's a mystery: Constructions of the future and 'un/becoming' amongst 'early' and 'late' career academics. *Int. Stud. Soc. Educ.* **2018**, *27*, 333–351. [CrossRef]
42. Junius-Walker, U.; Onder, G.; Soleymani, D.; Wiese, B.; Albaina, O.; Bernabei, R.; Marzetti, E. The essence of frailty: A systematic review and qualitative synthesis on frailty concepts and definitions. *Eur. J. Int. Med.* **2018**, *56*, 3–10. [CrossRef] [PubMed]
43. Gwyther, H.; Shaw, R.; Dauden, E.A.J.; D'Avanzo, B.; Kurpas, D.; Bujnowska-Fedak, M.; Holland, C. Understanding frailty: A qualitative study of European healthcare policy-makers' approaches to frailty screening and management. *BMJ Open* **2018**, *8*, e018653. [CrossRef] [PubMed]
44. Jones, S. Academic leadership. In *Pedagogic Frailty and Resilience in the University*; Kinchin, I.M., Winstone, N.E., Eds.; Sense Publishers: Rotterdam, The Netherlands, 2017; pp. 163–178.
45. Broglia, E.; Millings, A.; Barkham, M. Challenges to addressing student mental health in embedded counselling services: A survey of UK higher and further education institutions. *Br. J. Guid. Couns.* **2017**, *46*, 441–455. [CrossRef]
46. Kitzrow, M.A. The mental health needs of today's college students: Challenges and recommendations. *NASPA J.* **2003**, *41*, 167–181. [CrossRef]
47. Kutcher, S.; Wei, Y.; Coniglio, C. Mental health literacy: Past, present and future. *Can. J. Psychiatry* **2016**, *61*, 154–158. [CrossRef]

48. Gorczynski, P.; Sims-Schouten, W.; Hill, D.; Wilson, J.C. Examining mental health literacy, help seeking behaviours, and mental health outcomes in UK university students. *J. Ment. Health Train. Educ. Pract.* **2017**, *12*, 111–120. [CrossRef]
49. Kenny, J. Re-empowering academics in a corporate culture: An exploration of workload and performativity in a university. *High. Educ.* **2018**, *75*, 365–380. [CrossRef]
50. Taylor, C.A.; Harris-Evans, J. Reconceptualising transition to Higher Education with Deleuze and Guattari. *Stud. High. Educ.* **2018**, *43*, 1254–1267. [CrossRef]
51. Deleuze, G.; Guattari, F. *A Thousand Plateaus: Capitalism and Schizophrenia*; Massumi, B., Ed.; Athlone Press: London, UK, 2004.
52. Buta, B.; Leder, D.; Miller, R.; Schoenborn, N.L.; Green, A.R.; Varadham, R. The use of figurative language to describe frailty in older adults. *J. Frailty Aging* **2018**, *7*, 127–133. [PubMed]
53. Kinchin, I.M.; Cabot, L.B. Reconsidering the dimensions of expertise: From linear stages towards dual processing. *Lond. Rev. Educ.* **2010**, *8*, 153–166. [CrossRef]
54. Aguiar, J.G.; Thumser, A.E.; Bailey, S.G.; Trinder, S.L.; Bailey, I.; Evans, D.L.; Kinchin, I.M. Scaffolding a collaborative process through concept mapping: A case study on faculty development. *PSU Res. Rev.* **2019**. [CrossRef]
55. Gravett, K.; Yakovchuk, N.; Kinchin, I.M. *Enhancing Student-Centred Teaching in Higher Education: The Landscape of Student-Staff Partnerships*; Palgrave Macmillan: London, UK, 2019.
56. Matthews, K.E.; Dwyer, A.; Russell, S.; Enright, E. It is a complicated thing: leaders' conceptions of students as partners in the neoliberal university. *Stud. High. Educ.* **2018**, 1–12. [CrossRef]
57. Gravett, K.; Kinchin, I.M.; Winstone, N.E. 'More than customers': Conceptions of students as partners held by students, staff, and institutional leaders. *Stud. High. Educ.* **2019**, 1–14. [CrossRef]
58. Skilbeck, J.K.; Arthur, A.; Seymour, J. Making sense of frailty: An ethnographic study of the experience of older people living with complex health problems. *Int. J. Older People Nurs.* **2018**, *13*, e12172. [CrossRef]
59. Gravett, K.; Winstone, N.E. 'Feedback interpreters': The role of learning development professionals in facilitating university students' engagement with feedback. *Teach. High. Educ.* **2018**, 1–16. [CrossRef]
60. Barnett, L. Learning Development. In *Exploring Pedagogic Frailty and Resilience: Case Studies of Academic Narrative*; Kinchin, I.M., Winstone, N.E., Eds.; Brill: Leiden, The Netherlands, 2018; pp. 187–203.
61. Medland, E. Academic Development. In *Exploring Pedagogic Frailty and Resilience: Case Studies of Academic Narrative*; Kinchin, I.M., Winstone, N.E., Eds.; Brill: Leiden, The Netherlands, 2018; pp. 153–168.
62. Healey, M.; Jenkins, A. The role of academic developers in embedding high-impact undergraduate research and inquiry in mainstream higher education: Twenty years' reflection. *Int. J. Acad. Dev.* **2018**, *23*, 52–64. [CrossRef]
63. Kinchin, I.M.; Heron, M.; Hosein, A.; Lygo-Baker, S.N.; Medland, E.; Morley, D.; Winstone, N.E. Researcher-led academic development. *Int. J. Acad. Dev.* **2019**, *23*, 339–354. [CrossRef]
64. Jonas, W.B. Salutogenesis: The defining concept for a new healthcare system. *Glob. Adv. Health Med.* **2014**, *3*, 82–91. [CrossRef] [PubMed]
65. Barnacle, R.; Dall'Alba, G. Committed to learn: Student engagement and care in higher education. *High. Educ. Res. Dev.* **2017**, *36*, 1326–1338. [CrossRef]
66. Becker, C.M.; Glascoff, M.A.; Felts, W.M. Salutogenesis 30 Years Later: Where do we go from here? *Glob. J. Health Educ. Promot.* **2010**, *13*, 25–32.

Article

# The Role and Efficacy of Creative Imagination in Concept Formation: A Study of Variables for Children in Primary School

**Javier Gonzalez Garcia [1,*] and Tirtha Prasad Mukhopadhyay [2]**

[1] Department of Music and Performing Arts; University of Guanajuato, Guanajuato 36000, México
[2] Department of Art and Business, University of Guanajuato, Salamanca 36766, México
* Correspondence: jr2000x@yahoo.es

Received: 10 May 2019; Accepted: 30 May 2019; Published: 5 July 2019

**Abstract:** Children's creative imagination is tested through tasks involving narrative and drawing abilities for participants between the age of 8 and 12 years. The test determines the relative importance of 'narrative' against 'graphic' imagination in interpretive, problem-solving strategies, and also considers how such distinctive functions of the creative imagination could affect 'general' creativity of the child learner. Participants were chosen from designated primary schools in the state of Guanajuato, Mexico. The test on creativity complements facts from observational methodology in a population of mixed Castilian-speaking children. The name of the test is *Prueba de Imaginación Creativa Niños* (2008) or 'Test of Creative Imagination in Children', the Castilian acronym being PIC-N. It comprised four sub-tests: Three designed to evaluate narrative (verbal) creativity, and one for drawing (i.e., graphic) creativity. The first three 'exercises' in the suite indicates (a) fluency, (b) flexibility, and (c) originality in narrative representations, whereas the fourth indexes (d) graphic abilities of the child learner. Results suggest that creative imagination causes variations in specific aspects of creativity, like narrative and graphic improvisation, and also modifies 'general' creativity as understood from the perspective of a developmental psychology of learning abilities in growing children within the defined age group.

**Keywords:** creative imagination; drawing; storytelling; primary education; psychometric analysis

## 1. Introduction to Creativity

*Creative Imagination* appears to be active from an early age and is instrumental in learning and problem-solving strategies. Indeed, such imagination is one of the major ways in which children ascribe meaning and communicate information, attributing new meaning to objects they did not know previously, and to form new experiences from verbal stimuli, images, and also patterns of sound (which was not examined for this project). The archaic research on children's creativity was undertaken by Vygotsky [1], following Piaget [2], but in a manner that predicts the trajectory of the imagination for children's learning and interpretative abilities. Vygotsky's premises constitute some basic laws of developmental psychology for, more specifically, what he was to call "creativity", a term denoting psychomotor processes for producing both mental and physical representations in the material world. Also, 'creativity', as Vygotsky defined it, refers to a correlate of what he called 'imagination'. The latter, in fact, seems to have a more mental and even neural substratum. In this paper, we shall use the term 'creative imagination' to denote integrated neurobehavioral functions that employ imagination [3].

Vygotsky's basic proposition revealed how children could use creative imagination to make sense of this world. As such, the Vygotskian proposition lends itself to later speculations in psychology of children where importance has been laid on the methods of constructing narratives of imagination by children for strategies of interpretation and meaning creation [4]. What is of more immediate value for this research is the controversy regarding the intrinsic nature of creative imagination.

Is children's imagination manifested in organized reproductions of experience already available in real life as Piaget originally claimed, or did it have innate properties that manifested itself in various faculties, like verbal processing or visual memory? Harris' experiments, in this regard, use more of an active developmental angle to explain the process of creative imagination. Harris advocates multiple ontologies that suggest that the imagination functions as a cluster of discrete neurobehavioral faculties, like a multitasking function. Research on children's fantasy undertaken in the 1980s and 1990s underscore how imagination could control a child's understanding of corresponding aspects of reality [5–8]. Assuming that various aspects of the imaginative intelligence could be isolated and tested for their efficacy in the developmental process is perhaps of essence for this project. We consider the singular paradigm of creativity but we are generally inclined to show that facets of language, verbal processing, and visual-imaginative precepts may constitute separate enclaves in children's minds and that it is possible that one or other aspect may be instrumental in the execution of predefined tasks. Interestingly, it should also show that creativity progresses by means of both selective and synchronic organization of elements, and that practical engagement with such elements can lead to a general development for children's interpretative or learning abilities. Among other things, the test packet employed here explains functions of at least two important faculties. Furthermore, observation confirms that such practice could influence development on a long-term basis, for which other temporally spaced tests might be necessary.

Research here indicates that applications of the principle may be effective for teaching and learning at the elementary and middle school levels [9].

We intended to measure how verbal stimuli, generated from reading and storytelling, actually completes and transforms levels of knowledge in the child. We set imaginative tasks to stimulate this sense of meaning and interest in relation to real life problems and to explore why children's fantasy worlds are "so significant and interesting" and see how, as Egan said, we can use "what we learn from this process for educational purposes" [10]. Imagination is a "new psychological process" for the child. It represents a "specifically human form of conscious activity". Like all functions of knowledge, it arises originally from "action" [11]. Since creative imagination can be discovered in the products of creative life, especially of children, we tried to see how elements of representation, by story-telling or drawing, are incorporated in meaningful representations. The task of narrating or drawing objects involves analysis and reintegration, reproducing arrays, and objectivity in complex formations. Hence, the justification for the methodology adopted for the test.

## 2. The Quantitative Framework

There are disputes with the psychometric approach to studies on creativity and creative imagination. Many scholars doubt that it is possible to quantify something as delicate as creative imagination through standardized tests [12,13]. Consistently, in keeping with this line of thinking, Ausubel et al. [14] proposed that creativity was an 'innate capacity'; that is, a particularized and substantial capacity and born with the individual, but it would be difficult to develop it under influences of environment. There are some creative traits, however, that if developed, could turn to function as supportive instruments for the intellect and the personality of a child.

We demonstrate similarly that categorical re-inforcements might actually help children in our designated age group not only to deal with immediate instruments, but also learn to apply creative imagination to such instruments for better alternatives. If proven, intrinsic abilities within creative-imaginative modules may be shown to have positive effects children in the primary school segment. As we said, this proposition is not new in the literature, but our research could probably confirm the presence of variables in the project. We observed creativity in children to hypothesize if certain elements already singled out in the literature are also not instrumental in domain of the tasks for the experiment. We got to confirm that, among other suggested variables, creative imagination involves at least the following steps: (1) 'Encounter', since any process begins with the presentation of a tangible problem to the individual who engages with it [15]. (2) 'Incubation', during which an individual is continuously inspired by the reality of the problem [16]. (3) 'Enlightenment', corresponding to the results of the process [17]. (4) Self-induced evaluation and verification, in which the participant takes time to check the best options and ideas adopted [18]. We also endorse the profile of the creative subject in terms of nine aspects, among several recognized instruments realized in such authorities as Guilford [19] and Torrance [20]: These are originality, flexibility, production or fluency, elaboration, analysis, synthesis, mental opening, communication and sensitivity to problem, and level of inventiveness [21]. These factors are reduced to more computable categories for the experiment, as are outlined in the section on methodology. Finally, what remains to be said is that these factors, after careful extraction, are set aside specifically for verbal and visual creativity—the two more common sensory functions responsible for semantically valid constructs.

## 3. Existing Models of Creativity: Review and Suggestions

The evaluation of creativity is, thus, a hotly debated topic ranging from skepticism about the possibility of evaluation to asserting that creativity must be evaluated for purposes of improvement and instruction. De la Torre [22] is one author that bets on evaluation. He proposed that creativity requires inter and transdisciplinary approaches, since it has psychological, pedagogical, neurobiological, and sociological connotations [23]. The importance of evaluation of is also defended by authors such as Kaufman [24], who states that creativity is evaluated not only to measure the performance of an individual, but to promote the same type of creativity through stimuli and incentives.

In order to evaluate creativity, different instruments have been employed, all of which have taken into account the three major avenues to its understanding—namely, that of 'creative process' [25–27], and finally, that of quantifying the 'creative product' [28].

We should, thus, draw attention to these attempts in a systematic manner to follow the debate on how attempts have been made to quantify, use, and develop creative practices in real terms. First of all, studies on creative persons are usually aimed at identifying 'personality', 'motivation', 'intelligence', 'cognitive styles', and 'knowledge' that creative people possess or employ [29–31]. Most of these studies conclude that there are certain factors or personality traits that are often associated with high creativity in very different areas. These factors could be increased tolerance for ambiguity, willingness to take risk, self-regulation, interest etc. In this sense, the creative person is described as being more sensitive to problems and information gaps and seems to manifest a need to construct more hypotheses, to investigate and to evaluate problems [26]. Some personality traits appear to be more frequent in people who stand out for their artistic creativity and yet there are others that are more frequently visible in fields of scientific inventions [28].

The other approach to factorial analysis of creative activity has been to consider the conditions responsible for execution of novel tasks. The main models in this group attempts to describe mental operations or phases that a person goes through or experiences during the making of a creative product, such as exploration, incubation, insight, evaluation, etc. This approach also invites categorization of creative thinking in terms that suggest that it is different from non-creative thinking: Such as, for example, the extent to which creative thinking involves conscious, as distinct from unconscious processes, or to what extent creative thinking is the result of external effects such as luck, or of long-term efforts such as persistence and hard work. An example of this process-centered approach is the "Geneplore" model proposed by Finke, Ward, and Smith [31] in which they identify two phases in the creative process: A first of "generation", generating many possible ideas that can solve a problem and a second phase of "exploration", in which the different possible solutions are evaluated to select the most appropriate one. Another well-known approach is the one proposed by Csikszentmihaly [25], which introduces, as we well know, the notion of "flow". We do take cognizance of this model, as well the former one, with their insistence on personality and processual actions.

But the third existing model seems to be more viable, especially for children where personality is less complex, and products of creativity, such as drawing, are easier to deconstruct. Creativity is a fundamental aspect of cognitive development and is behaviorally manifested, and hence, its measurement depends on variables reflected in the immediate products of psychomotor functions, and this is perhaps more readily visible in children [32]. It is necessary to both evaluate and quantify the behavioral manifests of creativity in order to understand if there are relevant correlations of elements within constituents of those products.

So, as for this creative product model of creative imagination, scholars like Sternberg [33] and Kaufman [24] point out that most creative ideas are characterized by the presence of three components. First, that the creative idea must refer to something different, original, new, or innovative. Second, creative ideas must lead to a quality product. Third, creative ideas will have to be appropriate for the task or problem that is presented. In short, it could be said that a creative product must be novel, of recognizably high quality, and relevant. Studies focusing on the creative 'product' try to determine as to how to judge whether a particular work (poem, musical composition, drawing, etc.) is creative or not. In this sense, some researchers propose to use the consensual assessment technique while estimating the creativity of products or achievements of people or groups [13]. This technique involves judges or experts in an area of knowledge, assigning scores to real products such as a written composition, a poem, or a drawing. We would like to argue in favor of the fact that creativity should always be evaluated in such concrete domains. An evaluative-quantitative approach centered on the product has the potential of being more objective, although it would have the drawback of not allowing us to know how the subject reaches the end product. It also makes it possible to identify subjects that have been remarkably creative, but it does not make it easier to identify subjects whose creative potential are yet to be developed [34].

The fourth factor, the pressures of the environment, emphasizes how the social environment, the models to which one is exposed and the cultural values and attitudes that surround us contribute, sometimes to a great extent, to foment or inhibit our creativity.

## 4. Creativity Testing in Children

Torrance [35] has made great contributions to the study of creativity; he became interested in things that can be done before, during, and after a lesson to increase creative thinking. He indicated that creativity is a process that is expressed in changes discovered, is a capacity that can be developed and, in children, it is something that is confirmed through their productions, such as stories, fantasies, and drawings. In addition, he designed a test to evaluate the four basic skills that reflect creativity: Namely, fluency, flexibility, originality, and elaboration. In the most recent version of his test published in 2008, he proposed some changes to the proposed indicators: Fluidity here refers to the ability to produce a large number of ideas; originality now involves, for him, the ability to contribute ideas or solutions that are far from the obvious, common, or established; elaboration appears as the aptitude of detailing ideas; finally, again, titles for creativity refer in the latest version of the analysis to the ability to generate ideas that capture the essence of drawings and sensory reproductions. Torrance also speaks of the ability to generate original ideas, with intense images and details in addition to the stimulus.

The *Torrance's Creative Thinking Test* [36] is based on Guilford's theory of intellect [37] and is a useful tool for evaluating both quantitative and qualitative aspects of divergent thinking, especially creative products [38]. This instrument consists of a group of useful tests to evaluate the creative process as a whole and also the specific skills that define it. The figurative expression section of Form 1A evaluates the level of imagination in constructs like drawings and computes the products of the three following activities: (a) Composition, (b) finish, and (c) parallels. The definition of creativity is far from reaching an unanimity among members of the scientific community, but despite the diversity of perspectives, most authors agree that it is a complex and multifaceted construct. Davis and Wechsler [39] and Vendramini and Oakland [40] validate the Torrance test for the evaluation of divergent thinking; these authors recognize the accuracy and validity of the assumptions regarding the structure of the novel image. Most other similar studies report on the structural validity of the dimensions assessed in the subtests of this test (fluidity, flexibility, originality, and elaboration) and even establish high correlations between them. The works of Guilford [19] and Torrance [35] set a milestone in the study and evaluation of creativity based on psychometric and factorial perspectives.

Empirical evidence supporting the test proposed by Torrance [35] had its beginnings with the 1959 analyses of high school students where the three of the components of his test—fluidity, flexibility, and originality—were extensively confirmed and agreed upon as being the best predictors of creative achievement. Five years later, Torrance resent a questionnaire to the entire population that participated in the research of 1959 and found that all predictors of creativity were significant at a level of 0.01. In order to examine the relationship between the various indicators, a later study used Plucker's structural model in which it was argued that intelligence also has a positive effect on Torrance's test [41]. Increments in the four factors—fluidity, flexibility, originality, and elaboration—when evaluated by judgments of experts, reinforced the predictive validity of the test. In Plucker's research, data were examined under a Pearson product-moment correlation and a positive relationship was found between the indicators of creativity and the criteria of intelligence in creative achievement. Various analyses such as Plucker's have confirmed that the Torrance test is highly predictive of creative production in each of its indicators. Research of Terman and Oden (1959), Bloom (1985) [42], and Torrance (1993) with highly intelligent and talented individuals emphasized the critical role of personality factors, opportunity, experiences, and other environmental aspects that could play a role relevant to the development of creativity. However, it is very clear that there are other additional factors that can help or prevent the recognition of creative characteristics.

## 5. The Socio-Educational Context

Several authors such as Sánchez, García, and Valdés [43] have indicated that in a country like Mexico, there are no valid or reliable instruments of measuring children's creativity. Therefore, it was considered indispensable to validate Torrance's creative thinking test for a sample of Mexican students and to confirm psychometric properties of creativity in the population analyzed. Creative development provides a path of maturation in which creative activity manifests itself at different levels and in different ways. The present research aimed to analyze the incidence of imaginative strategies solely for the narrative and drawing tasks assigned to the children of the fifth grade in primary schools in Guanajuato, Mexico. The application consisted of a set of imaginative strategies in narrative and drawing tasks for an experimental group, and also for children not subject to the program, or the control group.

The assumptions of Torrance [35] and the later Torrance (2008) were considered as the starting hypotheses. Individuals who exhibit high levels of creativity have a greater potential to benefit from creative educational experiences and could be compared to groups of children with high and low levels of creativity [34] as is also evidenced in the research of Olveira, Ferrándiz, Ferrando, Sainz, and Prieto [44]. As for the only specific precedence, Soto and colleagues [45] already analyzed the construct validity of the Torrance creative thinking test with a sample of 500 students from a primary school located in an urban-marginal zone of the Iztapalapa delegation in the Federal District of Mexico.

Soto and his colleagues found that children with high scores showed an increase in graphic (visual) creativity and also originality, elaboration, fluency, better title construction, and closure. Children with low scores demonstrated increases in visual creativity and indicators of fluidity, originality, and titles. It is important to note that, although significant improvement in total creativity was achieved in both groups, a separate analysis was done for the group of highly creative children. Positive changes were observed in all the indicators, while in the group with low observed creativity, increase was recorded only in fluency, originality, and title-writing. Likewise, a decrease in the values of elaboration and closure was found for the same by Soto and his colleagues. Differences in individual parameters support the need of considering a variety of indicators for evaluating creativity, since the scores exhibited significant increases in both intragroup and intergroup creativity, as well as in each one of the indicators discussed. Different aspects of the same construct of creativity may show a certain independence from one another and this may be a desirable quality of the instruments of measurement.

Likewise, performance tests, questionnaires, self-reports, and tests involving subjective judgment have been used to assess creativity in children from other Latin American countries [23,46,47]. The tests of performance or skill that quantify the creative process, primarily of divergent thinking, are the tests that have focused most on the psychometric factors of creativity. In this sense, López-Martinez and Navarro-Lozano [47] talks about tests of divergent thinking that are elaborated from Guilford's SOI test. We also refer to Perez and Avila [48], whose tests are based on Guilford's multifactorial theory of intelligence, which states that creativity is not an independent dimension, but is integrated in contexts of many cognitive functions. Performance tests in Spanish made considerable use of versions of the PIC instrument. In addition to performance tests, the *PIC* has related assessment tools: Self-report, questionnaires, inventories, and scales that evaluate the characteristics of members in the sample. But these tests have limitations and are specific to the objective of research. Based on what has been said, this study aims to analyze if there is a statistical relationship between the scores obtained in narrative and drawing manifests of creative imagination and, unlike Ramirez or Lopez's study of additional variables, aims to be a comparative analysis of factors that are agreed upon in the literature. The *PIC-A* tests is based on the Guilford test battery, the turtle test, and latest versions of Torrance, and may be explained in detail for purposes of arriving at conclusions regarding merits of relative factors in creativity and also their synergic relationship to the broad question of how creative imagination functions as a system.

## 6. Methodology

The *PIC*, or *Prueba de Imaginación Creativa Niños* [49], was applied to count creativity for the experimental group (as well as the control group). The *PIC* may be translated as the 'Test of Creative Imagination in Children'. Generally, the *PIC* may be defined as a test of divergent thinking that evaluates creativity by examining how individuals use creative imagination in their creative representations. There are three versions of the *PIC*, for the levels of children *PIC-N*, adolescents *PIC-J*, and adults *PIC-A*, respectively, but all three versions have a similar structure of factorial measure for both verbal creativity and visual or graphic creativity.

Its application can be either individual or collective; in our case, the *PIC* was applied to participants individually. It also measures a 'general' creativity index for each participant. The scope of the experiment was restricted to school children between the third and sixth year of their primary education (with ages between 8 to 12 years). In terms of duration, the application could be variable, although for our experiment, a time-frame of approximately 40 min was chosen. For this duration, the *PIC* could be used to measure things like the ability to: (a) Formulate hypotheses (Exercise 1), (b) to think in an unconventional way (Exercise 2), (c) to exploit imagination and fantasy (Exercise 3), and (d) to employ, specifically, the graphic or imagistic imagination (Exercise 4). Results for the *PIC* are expressed in percentile for each variable studied and for each course of activity. Finally, we need to say that the materials for the test application was comprised of a test manual, an exercise copy, and a correction booklet.

## 7. Sample

The sample consisted of 300 children, 150 of them belonging to the experimental group (75 boys and 75 girls), and 150 corresponding to the control group (75 boys and 75 girls) from two public schools in Guanajuato, Mexico. The ages of the children ranged from 8 years to 12 years; students had to meet the following selection criteria: (a) Children could not have repeated or failed courses, (b) they could not have taken art classes (like drawing or painting) or literature classes, in which parents or guardians in charge might have influenced them in one of these two areas by demonstrating precepts of imaginative instruction and interpretation. These aforementioned variables, namely those of art or literature, were considered to restrict any prior intervention of factors that could have either affected or strengthened children's awareness or employment of creative imagination in some way or the other. Children who had this prior stimulus may have had higher scores. To evade the intervention of these variables, we had to make sure that the children population were equivalent in both the experimental and control groups.

## 8. Instruments

The *PIC* 2004, as we said, evaluates narrative creativity and graphic creativity to obtain a result of general creativity for the imagination of children between 8 and 12 years. The 4 "tests" or "exercises" are used to count four different facets of creativity originally suggested by Torrance (1966): Namely, fluidity, flexibility, elaboration, and originality, both in their narrative and graphic aspects. It offers scores on each of these facets and gives then a sum of both narrative and graphic aspects to render a total score of creativity. It may be characterized as a simple instrument, easy to apply or readjust with the help of guidelines given in the test manual. The term "game" is used on the front cover of the tests in order to minimize the impression of an evaluation or examination for participating subjects, and to obtain results of the application in an environment free from strict regulations since this encourages the ludic tendencies of creative subjects. The first game evaluates fluidity and flexibility with the help of an image in which children must write something by using given phrases to represent all the ideas that they consider may be passing through their minds. In the second, fluidity, flexibility, and originality are evaluated through a score for arrangement of one or several rubber tubes of different sizes. The third test poses an unexpected situation for the young participants, which is that they would need to imagine what would happen under implausible conditions, like in a situation in which a squirrel would suddenly turn into a dinosaur. Finally, four incomplete drawings are presented in the fourth game. Here, children had to complete each inchoate figure by drawing out something no one has thought of before.

A pre-test and post-test were applied, for evaluating fluidity, flexibility, originality, and elaboration according to the following criteria:

- Fluency: Evaluated in games 1, 2, and 3. One point is registered for each proposed idea. Repeated and non-pertinent answers are eliminated. To obtain score the total points for the individual games (1, 2, and 3) has to be added and placed in the corresponding box.
- Flexibility: This is evaluated in games 1, 2, and 3. Again, one point is registered for each different category. The number of categories is counted. To obtain the total, points from individual games (1, 2, and 3) must be added and then placed in the corresponding box.
- Originality: This is evaluated in games 2, 3, and 4. Scores are obtained by multiplying the coefficient that appears on the right, to obtain the total; similarly, all points must be added and placed in the corresponding box. This score is given on a range of 0 to 3. In game 4, originality is evaluated graphically by comparing the given answers of each drawing with the tables of originality that are provided. A score of 0, 1, 2, or 3 has to be assigned depending on the matching statistical frequency of the response.
- Elaboration: The answer given to each drawing was assigned a score of 0 to 2 according to the amount of details in it.

The application of the test was carried out individually, since it was assumed that the creative imagination is a process that is strengthened individually.

## 9. Description of Experimental Procedure

To achieve the objectives set out in this research, which is to determine how creative imagination augments learning and interpretive abilities, we examined an experimental group with pretest-and-post-test measurement, with a control group.

This measurement was made to determine the effects of a set of imaginative strategies (independent variables) on the dependent variable (creative imagination). In order to increase the internal validity of the study, intervening variables were controlled, thus excluding variables such as training and repetition.

## 10. A Detailed Report of the Pre-Test Activity

(1) A pre-test procedure was carried out, in order to ascertain the initial level of creative imagination in participants of both groups. This was done on an individual basis, in the same school, and on the same day.

Teachers and parents were informed of the objectives of the study and were solicited for confidentiality. The families of the students gave authorization to carry out the study.

Prior to the application of the program, the group classes that were to constitute the experimental group where the program was applied, and the group that was to constitute the control group, in which only the *PIC* was applied, were randomly selected. The test was applied before and after the development of the program in the experimental groups. The program was carried out in two centers.

The *PIC* pass is made during class hours and in a classroom where students are alone and separated from their peers. It is a sunny classroom, well ventilated, and noise-free, following the original instructions of the test manuals [49]. The application is made by the school counselor of the center.

Each child received a copy of the application booklet. During the application, the children had a varied material to paint (pens, waxes, markers, pencil sharpeners, erasers, etc.). The instructions were adapted to the children's age and comprehension skills. The applications were collective respecting the distribution by classrooms, in a single session.

The test is applied at the beginning (pretest) and at the end (post-test) of the program development to both the control group and the experimental group.

(2) Subsequently, it was applied to the strategies of the creative imagination that contained new activities in the experimental group. The first strategy for increasing fluency and flexibility required four activities (Guilford circles, mystery problems, brainstorming, other uses of everyday objects). The other strategy for incremental originality and elaboration was shaped by five activities (verbal analogies, relaxation and imagination, the best animal in the world, improvement of designs, and the machine without use), which sought to increase the participants' creative imagination through indicators raised by Guilford (1950) and confirmed and developed by Torrance. Once again, these are fluency, flexibility, originality, and elaboration.

## 11. Requirements and Exercises

The test consists of the accomplishment of exercises of complementation of drawings or graphs previously designed and proposed so that the students complete them; they must provide all the necessary ideas to make the drawing interesting. The test evaluates the creativity from the use that the individual makes of his imagination. It consists of four exercises: The first three assess verbal or narrative creativity and the fourth evaluates graphic creativity.

**Exercise 1**. In this exercise, from a situation that is reflected in a drawing, the individual has to write everything that could be happening in the scene. The presented stimulus varies according to the version of the *PIC* in question: In the *PIC-N*, a boy opening a chest; in the *PIC-J*, a boy and a girl in a lake; and in the *PIC-A*, an ambiguous scene is presented in the street, in which several characters appear. This game allows the person to express their curiosity and imagination and has been included to explore the ability to formulate hypotheses and think in terms of the possible. The test allows the expression of curiosity and speculative attitude; the ability to go beyond the information provided by the stimulus, by posing different possibilities with respect to what is imagined to occur on the scene.

**Exercise 2**. This exercise consists of a test, in which a list of possible uses of an object must be elaborated, according to Artola et al. [49]. In this case, it is: "Uses of a rubber tube"; in this exercise or subtest, the stimulus presented is the same in *PIC-N* as in *PIC-J* and *PIC-A*. This test is included as a measure of the ability of individuals to free their mind and think in an unconventional way; allows the evaluation of the "redefinition" of problems, that is, the ability to find uses, functions, and applications different from the usual ones; to quicken the mind; and to offer new interpretations or meanings to familiar objects, to give them a new use or meaning. Subjects can use the number and size of tubes they want. As in the first exercise, there is a list from 1 to 38, and it starts with an example.

**Exercise 3.** The exercise proposes subjecting subjects to unlikely situations. The situation presented varies according to the version of the *PIC*. In *PIC-N*, the situation is as follows: "Imagine what would happen if each squirrel suddenly became a dinosaur"; *in* the *PIC-J*, the situation is: "Imagine what would happen if the ground were elastic"; *in* the *PIC-A*: "Imagine what would happen if we did not stop remembering". The exercise evaluates the fantasy aspect of the imagination. This way of thinking seems very important in creative behavior. This exercise identifies the capacity for fantasy and the ability to handle unconventional ideas, which the subject probably would not dare to express in more serious situations, as well as openness and receptivity to novel situations. It is interesting how the test allows for the evaluation of the ability of "penetration" of the subject or ability to delve into experiences. Some of the consequences of the presented situation are obvious and simple to discover, while others, more remote, require a deeper study of the matter.

**Exercise 4.** This exercise is a graphic imagination test, inspired by the items of the Torrance Test (1966), according to Artola et al. [49]; in it, the subject must complete four drawings from a given stroke and put a title to each one of them. According to the authors, the incomplete figures used in game four have been selected, after presenting several figures to a sample of people considered as very creative (included in a program for gifted individuals), selecting the four that are most suggestive for them. The only premise, before starting the test, is to ask them to try to draw a picture that no other person could imagine; equally, the answers must give all the ideas necessary to make the drawing interesting.

## 12. Post-Test Assessments

(1). Nine activities were planned with three sessions per week. These activities took into account everyday situations and objects that allowed participant children to express their imagination through free expressions and reflecting them in a concrete way. Especially, narrative and graphic aspects were stimulated given that both aspects are exercised in different areas of activity in school.

(2). After the intervention, the post-test evaluation was performed for both groups individually. The objective was to analyze the impact of imaginative strategies in the narrative and the drawing reflected in the context of a significant increase of score in the experimental group.

(3). After the post-test evaluation, an open interview with the teachers was used to compare the results obtained and to develop a more in-depth discussion of the results of the intervention. (Appendix A).

## 13. Results

Exercises 1–4, in their individual and combined applications, yielded the following results for the four conditions underlined by Torrance [35] and developed in contexts of schoolchildren in Mexico by Ramirez and colleagues [45], and also for investigative paradigm among primary school children in Chile by Lubart [38]; Lopéz and Navarro [23] indicate a $p$-value < 0.01 for general creativity for a comparative analysis of experiment and control groups in these categories.

Through Tables 1 and 2 we can observe the evolution of results before and after the program of nine activities for creative development. Through the evolution of means and standard deviations, the impact of the imaginative strategies in the narrative and the reflected drawing is observed in the context of a significant increase in the score in the experimental group.

**Table 1.** Mean and standard deviation calculated on EXCEL for pre-test variables of creativity for the individual components suggested in the literature following Torrance (1966).

| Variable | Group | Median | Standard Deviation |
|---|---|---|---|
| Narrative Fluidity | Experimental | 19.3 | 3.42 |
| | Control | 20.2 | 3.49 |
| Narrative Flexibility | Experimental | 17.6 | 3.31 |
| | Control | 18.6 | 3.43 |
| Narrative Originality | Experimental | 11.2 | 2.91 |
| | Control | 11.5 | 2.87 |
| Graphic Originality | Experimental | 6.1 | 2.03 |
| | Control | 5.9 | 2.19 |
| Special Details | Experimental | 0.2 | 0.12 |
| | Control | 0.2 | 0.11 |
| Title | Experimental | 2.1 | 1.03 |
| | Control | 2.2 | 1.05 |
| Color and Shade | Experimental | 0.3 | 0.07 |
| | Control | 0.4 | 0.13 |
| Elaboration | Experimental | 2 | 0.4 |
| | Control | 1.8 | 0.45 |

**Table 2.** Standard deviation in post-test variables for individual Torrance components of narrative and graphic creativity for individually tested experimental and control groups.

| Variable | Group | Mean | Deviation |
|---|---|---|---|
| Narrative Fluidity | Experimental | 25.4 | 3.82 |
| | Control | 23.5 | 3.19 |
| Narrative Flexibility | Experimental | 21.2 | 3.05 |
| | Control | 19.1 | 3.13 |
| Narrative Originality | Experimental | 15.3 | 3.41 |
| | Control | 14.5 | 3.07 |
| Graphic Originality | Experimental | 6.1 | 1.92 |
| | Control | 5.9 | 2.01 |
| Special Details | Experimental | 0.2 | 0.24 |
| | Control | 0.2 | 0.15 |
| Title | Experimental | 2.4 | 0.93 |
| | Control | 2.3 | 1.12 |
| Color and Shade | Experimental | 0.3 | 0.07 |
| | Control | 0.4 | 0.13 |
| Elaboration | Experimental | 1.6 | 0.41 |
| | Control | 1.7 | 0.42 |

Table 3 shows results of inferential analysis obtained from the differences following interventions related to each one of the variables, namely of verbal and figurative creativity. In the first column, the mean change score is presented; in the second, standard deviation; while in the third column, the significance of this difference is presented. In the follow-up test, the test scores of the TCTT and the PIC-J are not comparable to each other at intra-group level because they are measured on different scales of measurement. For this reason, only the differences observed at the intergroup level are analyzed (Table 3).

**Table 3.** Mean, standard deviation, and ANOVA results of intergroup differences in the creativity of the follow-up test (*PIC-N*).

| Variable | Group | M | SD | P |
|---|---|---|---|---|
| **Narrative Creativity** | Experimental | 52.5 | 7.02 | 0.009 ** |
| | Control | 46.3 | 6.51 | |
| Narrative Fluidity | Experimental | 27.1 | 3.82 | 0.072 * |
| | Control | 26.6 | 3.19 | |
| Narrative Flexibility | Experimental | 15.4 | 3.05 | 0.0001 *** |
| | Control | 15.9 | 3.13 | |
| Narrative Originality | Experimental | 11.8 | 3.41 | 0.005 ** |
| | Control | 12.2 | 3.07 | |
| **Graphic Creativity** | Experimental | 13.46 | 4.42 | 0.001 * |
| | Control | 12.12 | 3.97 | |
| Graphic Originality | Experimental | 5.6 | 1.92 | 0.72 |
| | Control | 5.5 | 2.01 | |
| Special Details | Experimental | 0.17 | 0.05 | 0.002 * |
| | Control | 0.11 | 0.05 | |
| Title | Experimental | 2 | 0.93 | 0.42 * |
| | Control | 2.2 | 1.12 | |
| Color and Shade | Experimental | 2.4 | 0.07 | 0.002 * |
| | Control | 2.2 | 0.13 | |
| Elaboration | Experimental | 2.4 | 0.41 | |
| | Control | 2.3 | 0.42 | |
| General Creativity | Experimental | 65.4 | 7.88 | 0.005 ** |
| | Control | 60.3 | 6.62 | |

Note: *** $p < 0.001$. ** $p < 0.01$ * $p < 0.05$.

The results obtained from the narrative and graphic creativity in the monitoring test at the nomothetic level show significant differences between the EG and the GC in most of the creativity variables (Table 3). Specifically, the development of GE creativity is significantly greater than that of CG in narrative flexibility, special details and graphic creativity ($p < 0.001$); in the elaboration, the narrative creativity, the graphic creativity and in the general creativity ($p < 0.01$) and; in graphic originality ($p < 0.05$). The development of narrative fluency and the title of the GE is greater than that of the CG, but the difference between one and the other is not significant ($p > 0.05$).

A comparison of means was carried out using the Student's *t*-statistic, in order to evaluate the level of significance achieved after applying the strategies to the control group. Similarly, Pearson correlation and linear regression between each indicator were established, and the overall result for creative imagination of the experimental group was determined to verify the degree of association between them. For this analysis, we used the statistical program SPSS for Windows. The following graphs show the results according to the following criteria. In the X axis, there are six ranges that contain the direct scores distributed in intervals: Where the range of the number of is greater, the greater the score. The Y-axis represents the relative frequencies, which indicate the percentages of the absolute frequencies of subjects grouped in each interval. Subsequently, the level of significance of the totals in both groups are analyzed with the Student's T, which is established from the midpoint of the six intervals, in which the direct scores obtained by the subjects are distributed. Finally, we establish the Pearson correlation in the experimental group between the indicators of the narrative and graphic creativity in front of the general creativity as evidence of influence. The following is a summary of the results of the open research.

In Figure 1, the narrative creativity dimension, results were obtained in the first four ranks for the control group, with a higher percentage in the third. In the experimental group, however, no results appear in the first column, and show an increase in the fourth column, reaching maximum representation in the last one. This points to an increase in scores for the experimental group as opposed to the control.

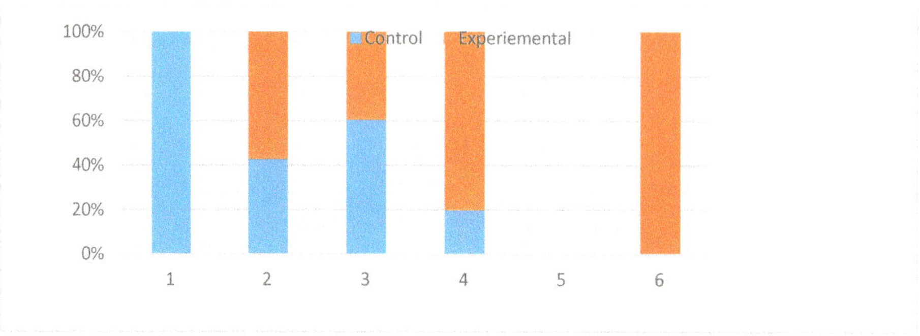


**Figure 1.** Narrative Creativity.

As shown in Figure 2, in the context of graphic or image-based creativity, the experimental group achieved its highest performance in the third range, followed by in the fourth one where its highest percentages are noted. This, as well as what we see is the fact that it obtained high percentages in the last two columns, unlike as in the control group, shows its highest percentages in the first three columns, standing out in the second, and without showing results in the last two.

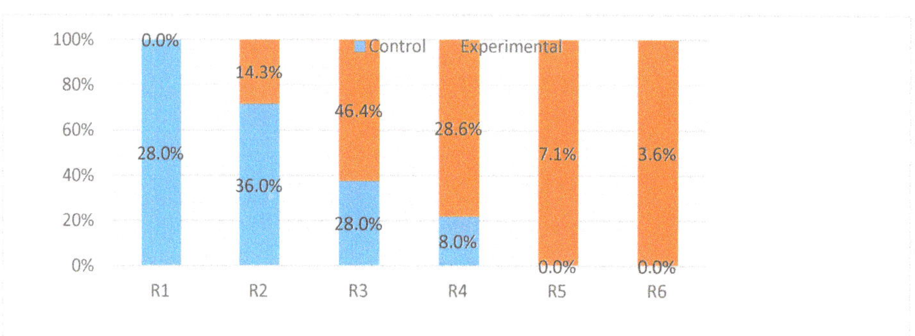


**Figure 2.** Graphic creativity (post-test).

As for general creativity (Figure 3), the control group scored in the first three ranges with their highest score in the third, unlike the experimental group that showed results from the second to the sixth range, but with the exception of the fifth.

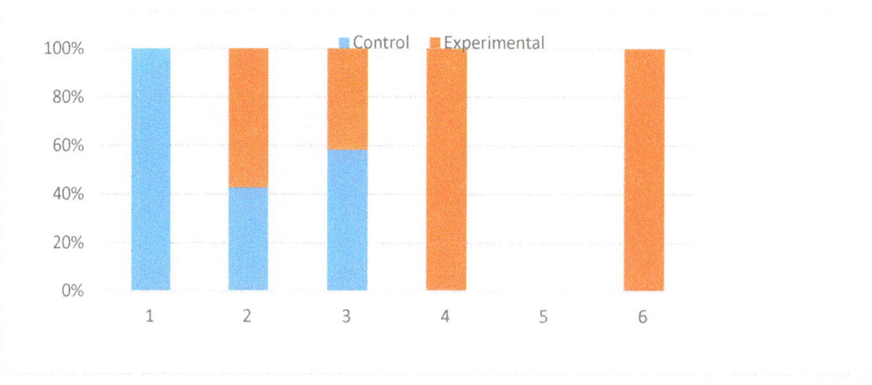


**Figure 3.** General creativity post-test.

As observed, the correlation is significant at the 0.01 level between the narrative dimension of the test and that of overall or general creative imagination. This result is explained by the fact that the strategies designed have a high component of activities related to narrative representation: Seven activities out of nine focus on narrative creativity. This means that the component that had the highest incidence in relation to the PIC test was narrative creativity. It seems that narrative creativity has a high correlation with fluidity, flexibility, and narrative originality (see Table 4).

**Table 4.** Correlation Creativity Narrative to General Creativity.

| | | NARRATIVE | GENERAL |
|---|---|---|---|
| NARRATIVE | Pearson Correlation | 1 | 0.994(**) |
| | Sig. (bilateral) | - | 0.000 |
| | N | 300 | 300 |
| GENERAL | Pearson Correlation | 0.994(**) | 1 |
| | Sig. (bilateral) | 0.000 | - |
| | N | 300 | 300 |

Note: *** $p < 0.001$. ** $p < 0.01$ * $p < 0.05$.

Unlike the previous result, the graphic dimension does not have a high correlation with a general creative imagination scale (Table 5). The results suggest that graphic originality and graphic design do not correlate more closely with total overall creativity (Appendix A). This means that the items of the PIC test, as well as the strategies employed, support the changes of creativity in terms of narrative rather than graphic creativity activity. However, as suggested in the discussion, the experiment demonstrates evidence of the importance of writing and narrative in general creative imagination.

**Table 5.** Correlation of graphic creativity to general creativity.

| CREATIVITY | | GRAFIC | GENERAL |
|---|---|---|---|
| GRAPHIC | Pearson Correlation | 1 | 0.135 |
| | Sig. (bilateral) | - | 0.493 |
| | N | 300 | 300 |
| GENERAL | Pearson Correlation | 0.135 | 1 |
| | Sig. (bilateral) | 0.493 | - |
| | N | 300 | 300 |

## 14. Post Test Qualitative Assessment

It is important to note that a complete evaluation of children's creative capacities requires attending to other environmental, academic, familial, and social factors that may influence the overall development of the child. This implies including qualitative information from observation scales performed by teachers or from interviews with children, parents, and teachers, or taking quantitative measures of attitudes and interests, peer nominations, teacher nominations, supervisor evaluations, product judgments, self-reports of activities, or creative achievements [50]. Taking as reference the Renzulli–Smith student characteristics rating scale and the Monterde primary school pupil observation scale, we followed up on open interviews and discussion groups with tutors and teachers that attended the performance of the sample of students. The objectives were several: To have a closer look, although approximately, at their habits outside the school; inserting reflections on creativity in a wider framework of knowledge of the people who contribute to the children's formation and on the places that host them. We met and listened to the people who lived with the sample of children, as a way to establish a relationship of mutual trust and respect and to approach the peculiar reality of schools. Interviews were conducted on the same school day, in which tests were applied to children, during recess or at a teacher-free time.

Tutors described boys and girls as generally restless and cheerful, and healthy in the broadest sense of the word. They defined them also as very autonomous in their daily chores. Most children come and go by themselves to school, are accustomed to go shopping from the age of seven, have their gang of friends with whom they go out, and know to organize their games.

In the opinion of the tutors, the areas in which children showed more creativity are organization of games using resources offered by the environment, the resolution of conflicts between themselves, and their crafts. The tutors agree that the environment in which they live offers endless resources for the promotion of their creativity. This, in addition to the freedom granted by their parents, helps them to become more aware of their skills and interests. Tutors considered autonomy as a fundamental aspect of the development of creativity. In their environment, children moved with confidence and responsibility; outside of it, during cultural visits for example, they seemed to lose their safety and sometimes appeared disoriented. In this case, the difference between those who are accustomed to going out with the family and moving in different contexts and those who are not, were noted. In general, they do not show special fears but are cautious and respectful of the unknown. There are no evident problems of discipline. Children manifest in general self-control and respect towards the school environment and the tutors.

The group of teachers who either attended or took care of the children reported positive attitude and motivation after the experience generated by this research project. Regardless of the disciplinary field in which they are found, the after-effects of artistic education can be adjusted to all educational processes. A majority of the teachers agreed that transdisciplinary performances were generally conducive to civic and ethical education, English, mathematics, and Spanish. A second group endorsed transdisciplinary plastic arts education for better development of precepts in civic formation and ethics, mathematics, and natural sciences. Likewise, this group of teachers expressed that they observed creative thinking skills in their fourth-grade students, which are similar to those that formed part of the conceptual framework of our research, especially to fluidity of thinking, originality, and elaboration. The rest of the qualities stated by the teachers have little relation to the qualities that reflect creative thinking abilities addressed in the conceptual framework of this research.

Most teachers agree that creativity does not consist of the perfect use of a technique to perform an artistic work, because they recognize the need of creativity to solve problems of daily life as well. Creative attitudes are often observed in students, in formats such as socializing and acquainting with new and unpublished objects or creations, followed by indications that they relate to knowledge from different fields to express their opinion on a topic. It includes production of novel hypotheses and questions on introductory topics, in the same way they express gracious analogies about everyday situations. It is logical to assume that many teachers find it very difficult to drive and follow up on experiences generating creative thinking skills in their educational plans if they themselves had no chance to experience them during their training processes in their respective training faculties, i.e., it is not enough to learn from the concept of memory or copy the characteristics of such thinking. It is necessary to make the activity existential [44].

Only two teachers were trained in concrete subjects of education and artistic appreciation, and a great majority are unaware that children could be educated through the creative imagination. Moreover, those who know him do not know how to do it. This awareness should lead to teachers using innovative strategies in the classroom, and to believe in education, because children can be educated in other ways.

## 15. Discussion of Results

The implementation of the set of strategies confirmed the role of creative imagination in generation of intelligible narratives and a smaller degree of its presence in graphic or imagistic solutions. While selecting the activities that formed the strategies for the increase of the creative imagination, four indicators were followed: Fluidity, flexibility, originality, and elaboration. Activities should involve criteria for the strengthening of each indicator, which, as a whole, would allow development of the creative imagination.

Children could have unique ideas that could be positively valued and could arguably provide them the confidence to freely express their ideas without the fear of being told that they are wrong.

Children who received the treatment showed receptivity and seemed to enjoy the activities. Creative imagination is a capacity that can be enhanced through the implementation of relevant strategies. Considering that both groups were equivalent with respect to the results of the pre-test, it was clearly noticed that the experimental group, while receiving the treatment, reached significantly higher levels of score in general creativity. Achievement of this increment means fluidity, flexibility, and originality are causative factors for the process. Indicators of narrative creativity demonstrated a high correlation compared to general creativity. This means that narrative indicators are relevant for the strengthening of the creative imagination. On the other hand, when originality and graphic activities are compared, indicators did not appear to reach superior significance levels, but demonstrated low positive correlations.

This result, which is still provisional, allows us to believe that despite availability of novel resources in direct acts of manufacture (involving things like unused machines, fantasy animals, verbal analogies, improved designs), there is no high impact of graphic or visually concomitant imagination on general creativity. If this result is extended to teaching scenarios in preschool and middle school children's education, it is possible to suggest that perhaps many graphic or drawing activities, which the teacher asks children to do, have less than expected results or are perhaps redundant in the long run.

Neither is the association of drawing directly relevant to creative imagination. These results suggest that creative imagination is not an act of *reproduction* [49,51], but a mental function that involves elaboration and styles of thought [52], and that part of concrete images, when strengthened with external stimuli—which, in this case, is represented by the given set of imaginative exercise—allows production of something novel, something that was reflected in specific actions and involved activities of an analytical character like writing.

## 16. Conclusions

The *PIC* appears to be a suitable instrument for detecting variations in general creativity. The results point out that not only can there be variations in specific aspects of creativity, but also that general creativity can be modified stochastically or otherwise, thus challenging the theoretical consideration of Ausubel et al. (1998), who proposed that multiple intrinsic or extrinsic factors, in their isolated or combined engagements, do not have a greater incidence on creativity. Though global in its cross-cultural applications, the proposal for incorporating imaginary worlds in children's curricula for children in Latin American contexts, such as Bronstein or Bruner suggested, may not be an ineffectual contingency.

The results converge with some of the objectives of the research. The implementation of a program that develops divergent thinking will favorably influence the creativity of students. It shows that a creativity development program represents an instrument for teachers and children that provides tasks and materials with which to rehearse a variety of ways to express their creative potential. These results also suggest that the creative imagination is not an act of representation or reproduction, but a mental function that involves thought, which starts from concrete images, and is strengthened with external stimuli: In this case, the process is borne out in the set of imaginative strategies that allowed us to effectuate something new. The effects were reflected in specific actions that involved analytical activities such as writing. Finally, it should be noted that the *PIC* proved to be an adequate instrument for measurement of variations in general creativity. Even though no pedagogical adaptations have been made in the Mexican social context, this study offers evidence that allows its use to be relevant in educational practice.

**Author Contributions:** Conceptualization, J.G.G. methodology, J.G.G. and T.P.M.; validation J.G.G. and T.P.M.; formal analysis, J.G.G. and T.P.M.; investigation, J.G.G.; resources, J.G.G.; data healing, J.G.G. and T.P.M.; writing-preparation of the original draft, J.G.G. and T.P.M.; writing-revision and editing, J.G.G. and T.P.M.; visualization, J.G.G.; supervision, J.G.G. and T.P.M. project management, J.G.G.

**Funding:** This investigation did not receive external financing.

**Conflicts of Interest:** The authors declare no conflict of interest.

## Appendix A

**Table A1.** Correlation indicators for narrative, graphic, and general creativity.

| Indicators | | Fluidity | Flexib. | Orignlty | Narrative | Graphic Orignlty. | Elabor Graph. | Graphic | General |
|---|---|---|---|---|---|---|---|---|---|
| FLUIDITY | Pearson Correlation | 1 | 0.857(**) | 0.832(**) | 0.967(**) | −0.041 | 0.164 | 0.071 | 0.963(**) |
| | Sig. (bilateral) | 0.000 | 0.000 | 0.000 | 0.000 | 0.835 | 0.405 | 0.718 | 0.000 |
| | N | 300 | 300 | 300 | 300 | 300 | 300 | 300 | 300 |
| FLEXIBILITY | Pearson Correlation | 0.857(**) | 1 | 0.754(**) | 0.911(**) | −0.041 | 0.048 | −0.001 | 0.901(**) |
| | Sig. (bilateral) | 0.000 | - | 0.000 | 0.000 | 0.835 | 0.808 | 0.996 | 0.000 |
| | N | 300 | 300 | 300 | 300 | 300 | 300 | 300 | 300 |
| ORIGINALITY | Pearson Correlation | 0.832(**) | 0.754(**) | 1 | 0.922(**) | 0.024 | 0.021 | 0.031 | 0.918(**) |
| | Sig. (bilateral) | 0.000 | 0.000 | - | 0.000 | 0.904 | 0.914 | 0.875 | 0.000 |
| | N | 300 | 300 | 300 | 300 | 300 | 300 | 300 | 300 |
| TOTAL NARRATIVE | Pearson Correlation | 0.967(**) | 0.911(**) | 0.922(**) | 1 | −0.038 | 0.110 | 0.041 | 0.994(**) |
| | Sig. (bilateral) | 0.000 | 0.000 | 0.000 | - | 0.849 | 0.576 | 0.837 | 0.000 |
| | N | 300 | 300 | 300 | 300 | 300 | 300 | 300 | 300 |
| ORIGINALITY GRAPHIC | Pearson Correlation | −0.041 | −0.041 | 0.024 | −0.038 | 1 | 0.058 | 0.782(**) | 0.044 |
| | Sig. (bilateral) | 0.835 | 0.835 | 0.904 | 0.849 | - | 0.770 | 0.000 | 0.824 |
| | N | 300 | 300 | 300 | 300 | 300 | 300 | 300 | 300 |
| ELABORATION GRAPHIC | Pearson Correlation | 0.164 | 0.048 | 0.021 | 0.110 | 0.058 | 1 | 0.667(**) | 0.164 |
| | Sig. (bilateral) | 0.405 | 0.808 | 0.914 | 0.576 | 0.770 | - | 0.000 | 0.404 |
| | N | 300 | 300 | 300 | 300 | 300 | 300 | 300 | 300 |
| TOTAL GRAPHIC | Pearson Correlation | 0.071 | −0.001 | 0.031 | 0.041 | 0.782(**) | 0.667(**) | 1 | 0.135 |
| | Sig. (bilateral) | 0.718 | 0.996 | 0.875 | 0.837 | 0.000 | 0.000 | - | 0.493 |
| | N | 300 | 300 | 300 | 300 | 300 | 300 | 300 | 300 |
| CREATIVITY GENERAL | Pearson Correlation | 0.963(**) | 0.901(**) | 0.918(**) | 0.994(**) | 0.044 | 0.164 | 0.135 | 1 |
| | Sig. (bilateral) | 0.000 | 0.000 | 0.000 | 0.000 | 0.824 | 0.404 | 0.493 | - |
| | N | 300 | 300 | 300 | 300 | 300 | 300 | 300 | 300 |

Note: ** Correlation is significant at the 0.01 level (bilateral).

## References


1. Vygotsky, L.S. *La Imaginación y el Arte en la Infancia;* Akai: Barcelona, Spain, 1990.
2. Piaget, J. *Play, Dreams and Imitation in Childhood;* Morton: New York, NY, USA, 1962.
3. Lindqvist, G. Vygotsky's theory of creativity. *Creat. Res. J.* **2003**, *15*, 245–251.
4. Vygotsky, L.S. Imagination and creativity in childhood. *J. Russ. East. Eur. Psychol.* **2004**, *42*, 4–84. [CrossRef]
5. Morison, P.; Gardner, H. Dragons and dinosaurs: The child's capacity to differentiate fantasy from reality. *Child Dev.* **1978**, *49*, 642–648. [CrossRef]
6. Carey, S. Thinking and learning skills. In *Are Children Fundamentally Different Kinds of Thinkers than Adults?* Chipman, S., Segal, J.W., Glaser, R., Eds.; Erlbaum: Hillsdale, NJ, USA, 1985; Volume 2.
7. Siegler, R.S. *Emerging Minds: The Process of Change in Children's Thinking;* Oxford University Press: New York, NY, USA, 1996.
8. Stanovich, K.E. Reconceptualizing intelligence: Dysrationalia as an intuition pump. *Educ. Res.* **1994**, *22*, 5–21. [CrossRef]

9. Singer, D.G.; Singer, J.L. *The House of Make-Believe: Children's Play and the Developing Imagination*; Harvard University Press: Cambridge, MA, USA, 1990.
10. Egan, K. *Fantasía e Imaginación: Su Poder en la Enseñanza*; MEC y Morata: Madrid, Spain, 1994.
11. Vygotsky, L.S. *El Desarrollo de Los Procesos Psicológicos Superiors*; Crítica: Barcelona, Spain, 1989.
12. Hocevar, D. Measurement of creativity: Review and critique. *J. Personal. Assess.* **1981**, *45*, 450–464. [CrossRef]
13. Baer, J.; McKool, S.S. Assessing creativity using the consensual assessment technique. In *Handbook of Research on Assessment Technologies, Methods, and Applications in Higher Education*; IGI Global: Hershey, PA, USA, 2009; pp. 65–77.
14. Ausubel, D.P. Schemata, cognitive structure, and advance organizers: A reply to Anderson, Spiro, and Anderson. *Am. Educ. Res. J.* **1980**, *17*, 400–404. [CrossRef]
15. May, R. Creativity and encounter. *Am. J. Psychoanal.* **1964**, *24*, 39–43. [CrossRef]
16. Gilhooly, K.J.; Georgiou, G.; Devery, U. Incubation and creativity: Do something different. *Think. Reason.* **2013**, *19*, 137–149. [CrossRef]
17. Wikander, L.; Gustafsson, C.; Riis, U. (Eds.) *Enlightenment, Creativity and Education: Polities, Politics, Performances*; Springer Science & Business Media: New York, NY, USA, 2012; Volume 19.
18. Silvia, P.J.; Wigert, B.; Reiter-Plamon, R.; Kaufman, J.C. Assessing creativity with self-report scales: A review and empirical evaluation. *Psychol. Aesthet. Creat. Arts* **2012**, *6*, 19–34. [CrossRef]
19. Guilford, J.P. Creativity. *Am. Psychol.* **1950**, *5*, 444–445. [CrossRef]
20. Torrance, E.P. The nature of creativity as manifest in its testing. In *The Nature of Creativity: Contemporary Psychological Perspectives*; Cambridge University Press: New York, NY, USA, 1988; pp. 43–75.
21. Lee, Y.J. Effects of Divergent Thinking Training/Instructions on Torrance Tests of Creative Thinking and Creative Performance. Ph.D. Thesis, University of Tennessee, Knoxville, TN, USA, 2004.
22. De La Torre, T. Los cuatro puntos cardinales en la evaluación de la creatividad. In *Comprender y Evaluar la Creatividad*; Ediciones Aljibe: Málaga, Spain, 2006; pp. 143–154.
23. López-Martínez, O.; Navarro-Lozano, J. Creatividad e inteligencia: Un estudio en educación primaria. *Rev. Investig. Educ.* **2010**, *28*, 283–296.
24. Kaufman, J.C. Counting the muses: Development of the kaufman domains of creativity scale (K-DOCS). *Psychol. Aesthet. Creat. Arts* **2012**, *6*, 298–308. [CrossRef]
25. Csikszentmihalyi, M. Sociedad, cultura y persona: Una perspectiva sistémica de la creatividad. In *La Naturaleza de la Creatividad*; Sternberg, R.J., Ed.; Cambridge University Press: New York, NY, USA, 1988.
26. Mccraken, J.R. Design—The creative soul of technology. *Essent. Top. Technol. Educ.* **2009**, *1001*, 69–73.
27. Feist, G.J. The influence of personality on artistic and scientific creativity. In *Handbook of Creativity*; Cambridge University Press: New York, NY, USA, 1999; Volume 14, p. 273.
28. Kauffman, J.C.; Beghetto, R.A. Beyond big and little: The four C model of creativity. *Rev. Gen. Psychol.* **2009**, *13*, 1–12. [CrossRef]
29. Sternberg, R.J.; Lubart, T.I. *Desafiando a la Multitud: Cultivar la Creatividad en una Cultura de la Conformidad*; Free Press: New York, NY, USA, 1995.
30. Amabile, T.M. *Creativity and Innovation in Organizations*; Harvard Business School: Cambridge, MA, USA, 1996.
31. Baer, J.; Kaufman, J.C. Bridging generality and specificity: The amusement park theoretical (APT) model of creativity. *Roeper Rev.* **2005**, *27*, 158–163. [CrossRef]
32. Finke, R.A.; Ward, T.B.; Smith, S.M. *Creative Cognition: Theory, Research and Applications*, 1st ed.; MIT Press: Cambridge, MA, USA, 1996.
33. Sternberg, R. Assessment of gifted student a new millennium. *Learn. Individ. Differ.* **2010**, *20*, 327–336. [CrossRef]
34. Runco, M.A. Divergent thinking, creativity, and giftedness. *Gift. Child Q.* **1993**, *37*, 16–22. [CrossRef]
35. Torrance, E.P. *Torrance Tests of Creative Thinking: Norms-Technical Manual: Verbal Tests, Forms A and B: Figural Tests, Forms A and B*; Personal Press Incorporated: Wichita, KS, USA, 1966.
36. Torrance, E.P. *Research Review for the Torrance test of Creative Thinking Figural and Verbal Forms A and B*; Scholastic Testing Service: Bensenville, IL, USA, 2008.
37. Guilford, J.P. *The Nature of Human Intelligence*; Mc Graw Hill: New York, NY, USA, 1967.
38. Lubart, T. Creativity from a Cognitive Developmental Science Perspective. Available online: http://csjarchive.cogsci.rpi.edu/ (accessed on 15 August 2017).

39. Davis, G.A. Testing for creative potential. *Contemp. Educ. Psychol.* **1989**, *14*, 257–274. [CrossRef]
40. Wechsler, S.M.; Vendramini, C.M.M.; Oakland, T. Thinking and creative styles: A validity study. *Creat. Res. J.* **2012**, *24*, 235–242. [CrossRef]
41. Plucker, J.A. Is the proof in the pudding? Reanalyses of Torrance's (1958 to present) longitudinal data. *Creat. Res. J.* **1999**, *12*, 103–114. [CrossRef]
42. Bloom, B.S.; Sosniak, L.A. *Developing Talent in Young People*; Ballantine Books: New York, NY, USA, 1985.
43. Escobedo, P.A.S.; Mendoza, A.G.; Cuervo, A.A.V. Validez y confiabilidad de un instrumento para medir la creatividad en adolescentes. *Rev. Iberoam. Educ.* **2009**, *50*, 1–12.
44. Olveira, E.; Ferranndiz, C.; Ferrando, M.; Sainz, M.; Prieto, M. Test de pensamiento creativo de torrance (ttct): Elementos para la validez de constructo en adolescentes portugueses. *Psicothema* **2009**, *21*, 562–567.
45. Soto, B.I.C.; Ramirez, F.Z.; Tomasini, G.A. ¿Quiénes son los alumnos con aptitud sobresaliente? Análisis de diversas variables para su identificación/who are students with outstanding ability? Analysis of different variables for identification. *Actual. Investig. Educ.* **2014**, *14*, 479–511.
46. Zacatelco, F.; Chavez, B.I.; Gonzalez, A.; Acle, G. Validez de una prueba de creatividad: Estudio en una muestra de estudiantes mexicanos de educación primaria. *Rev. Intercont. Psicol. Educ.* **2013**, *15*, 141–155.
47. Lopez-Martin, O.; Navarro-Lozano, J. Influencia de una metodología creativa en el aula de primaria. *Eur. J. Educ. Psychol.* **2010**, *3*, 89–102. Available online: http://www.redalyc.org/articulo.oa?id=129313736007 (accessed on 6 September 2017). [CrossRef]
48. Perez, V.H.; Ávila, F. La unificación de criterios en torno a la medición del constructo creativo. *Cienc. Empresariales* **2014**, *23*, 16–28.
49. Artola, T.; Barraca, J. Creatividad e imaginación. Un nuevo instrumento de medida: La PIC. *EduPsykhé* **2004**, *3*, 73–93.
50. Zeng, Y. Environment-based design (EBD). In Proceedings of the ASME 2011 International Design Engineering Technical Conferences and Computers and Information in Engineering Conference, Washington, DC, USA, 28–31 August 2011; Volume 54860, pp. 237–250.
51. Menchen, F. Atrévete a ser creativo: Pasos para ser creativos. *Rev. Iberoam. Sobre Calid. Efic. Cambio Educ.* **2012**, *10*, 249–263.
52. Sternberg, R.J.; Grigorenko, E.L. Are cognitive styles still in style? *Am. Psychol.* **1997**, *52*, 700–712. [CrossRef]

*Article*

# Fundamental Issues of Concept Mapping Relevant to Discipline-Based Education: A Perspective of Manufacturing Engineering

**AMM Sharif Ullah**

Faculty of Engineering, Kitami Institute of Technology, 165 Koen-cho, Kitami 090-8507, Japan; ullah@mail.kitami-it.ac.jp

Received: 27 July 2019; Accepted: 26 August 2019; Published: 29 August 2019

**Abstract:** This article addresses some fundamental issues of concept mapping relevant to discipline-based education. The focus is on manufacturing knowledge representation from the viewpoints of both human and machine learning. The concept of new-generation manufacturing (Industry 4.0, smart manufacturing, and connected factory) necessitates learning factory (human learning) and human-cyber-physical systems (machine learning). Both learning factory and human-cyber-physical systems require semantic web-embedded dynamic knowledge bases, which are subjected to syntax (machine-to-machine communication), semantics (the meaning of the contents), and pragmatics (the preferences of individuals involved). This article argues that knowledge-aware concept mapping is a solution to create and analyze the semantic web-embedded dynamic knowledge bases for both human and machine learning. Accordingly, this article defines five types of knowledge, namely, analytic a priori knowledge, synthetic a priori knowledge, synthetic a posteriori knowledge, meaningful knowledge, and skeptic knowledge. These types of knowledge help find some rules and guidelines to create and analyze concept maps for the purposes human and machine learning. The presence of these types of knowledge is elucidated using a real-life manufacturing knowledge representation case. Their implications in learning manufacturing knowledge are also described. The outcomes of this article help install knowledge-aware concept maps for discipline-based education.

**Keywords:** concept map; learning; semantic web; knowledge representation; epistemology

---

## 1. Introduction

This article addresses some fundamental issues regarding concept mapping for discipline-based education. The focus is on manufacturing knowledge representation from the viewpoints of both human and machine learning, and the context is new-generation manufacturing (Industry 4.0, smart manufacturing, and connected factory).

Many authors have studied about how to learn science- and engineering-based subject matters in the early stage of formal education. Accordingly, it has been found that not only knowing about a subject matter, but also doing about it (arguing about scientific theories and findings, suggesting plausible solutions, and so on) can enhance learning [1–7]. This kind of education (knowing and doing simultaneously) requires integration between theoretical and real worlds following three phases, namely, investigating, evaluating, and developing explanations and solutions. As a result, the learning must be driven by the following activities: (1) Asking questions and defining problems, (2) developing and using models, (3) planning and carrying out investigations, (4) analyzing and interpreting data, (5) using mathematics and computational thinking, (6) constructing explanations and designing solutions, (7) engaging in argument from evidence, and (8) obtaining, evaluating, and communicating information [8]. Having said that, it might not be true that all learners are free from misconceptions [9–12]. Some efforts are required to help learners overcome misconceptions.

When a learner continues her/his science- and engineering-based education at the tertiary level, the abovementioned duality (knower–doer) intensifies due to some predefined educational objectives and outcomes (e.g., the educational objectives and outcomes of the Accreditation Board for Engineering and Technology (ABET) [13]).

Regardless of the level of formal education (school, undergraduate, or graduate level), some motivating factors drive an individual to become a knower and doer simultaneously. The author believes that two of the motivating factors are metacognition and meaningful learning. Metacognition (thinking about thinking) is a higher-order human cognition that allows individuals to monitor and redirect their thinking processes as needed [14,15]. Meaningful learning is perhaps a manifestation of metacognition that emotionally attaches an individual to a learning process, resulting in a concept map (a network of concepts) [16–20]. A remarkable feature of meaningful learning is that it integrates new concepts with existing ones.

Concept map-based education has earned a great deal of attention [20–25]. It has spread up to discipline-based education, including manufacturing engineering education [25]. As far as manufacturing engineering education is concerned, concept map-based learning is significant from the viewpoints of both human and machine learning. The significance of concept maps from the viewpoint of human learning in manufacturing evolves due to the advent of learning factory [26–28], whereas the significance of concept maps in manufacturing from the viewpoint of machine learning evolves due to the advent of new-generation manufacturing systems [29–31]. Thus, as far as the advancement of digital manufacturing is concerned, the construction process of concept maps has become an important issue. A few authors have researched the construction process of concept maps in general. Some of the noteworthy processes are as follows: Semantic gravity-driven concept mapping [21], process, material, automation, and shape universe-based concept mapping [25], focus-question-based concept mapping [32], and weighted concept induction-based concept mapping [33]. The remarkable thing is that the contents of a concept map (intended for human or machine learning) boil down to some propositions. These propositions can be categorized into some types of knowledge [21,31]. Therefore, knowledge-type-aware concept mapping is one of the effective processes of constructing concept maps. For this perspective, this article is written.

Accordingly, this article aims to provide more insights into the general categorization of knowledge and its representation using concept maps from the perspectives of manufacturing engineering education. Therefore, this article must describe the fundamental issues of manufacturing engineering knowledge and its ICT-based representation from the perspectives of human and machine learning. It must define the knowledge types from a domain-neutral perspective (i.e., epistemology). Lastly, it must elucidate how to accommodate the knowledge types into the learning activities in manufacturing engineering education through the formation of concept maps. Therefore, the rest of this article is organized as follows. Section 2 presents the fundamental issue underlying concept mapping in manufacturing engineering. Section 3 provides a knowledge classification method for organizing the manufacturing engineering-relevant contents for concept mapping. Section 4 analyzes a manufacturing engineering-relevant concept map using the proposed knowledge classification. Section 5 concludes this study.

## 2. Fundamental Issues Regarding Manufacturing Knowledge Representation

The previous section states that concept mapping in manufacturing is significant from both human and machine learning perspectives. The significance in terms of human learning evolves due to the advent of a concept called learning factory [26–28], whereas the significance in terms of machine learning evolves due to the advent of a concept called new-generation manufacturing systems [29–31]. On the other hand, the concept of learning factory is heavily linked to the concept of new-generation manufacturing systems. Thus, before eliciting the fundamental issues of manufacturing knowledge representation using concept maps, the relevant aspects of new-generation manufacturing systems must be elucidated. Accordingly, this section presents some of the fundamental issues relevant to knowledge representation that center around new-generation manufacturing systems.

Manufacturing (or production) systems have continuously been evolving under the influence of information and communication technology. As a result, a concept of new-generation manufacturing systems has been evolved, which is referred to as Industry 4.0, smart manufacturing, connected factory, Society 5.0, Made in China 2025, and alike [26–31,34–40]. The primary goal is to achieve an active collaboration among hardware devices (e.g., machine tools, robots, measuring instruments), software systems (CAD/CAM, ERP, and SCM systems), and human resources on a real-time basis by exchanging the required data, information, and knowledge [34–40]. For achieving this goal, a set of relevant technologies has been introduced, namely, human-cyber-physical systems, digital twins, and the Internet of things [26–31,34–40]. Numerous authors have studied these technologies. For example, Zheng et al. [31] reviewed Industry 4.0 and provided a system architecture where the data intensiveness of design, monitoring, machining, control, and scheduling are classified into four layers—namely, sensor and actuator deployment, data collection, big data analysis, and big data-driven decision-making. The layers are organized in the order of edge (where an activity occurs), fog (cyber-physical-human-integrated systems), and cloud (an ICT infrastructure for collecting and disseminating information from/to a wide range of stakeholders). Koren et al. [34] described how to modify the reconfigurable manufacturing systems in order to accommodate the functionalities of Industry 4.0 [29,32]. They have emphasized that the human-cyber-physical systems or the systems that reside in the fog (the mid-wear between edges and cloud) must be populated with the contents called digital twin (computable virtual abstraction of real objects, processes, and phenomena). Ullah [30,38] and Ghosh et al. [40] described that there are three kinds of digital twins, namely, object twin, process twin, and phenomenon twin. Among these twins, phenomenon twin is the most challenging twin to construct because the construction of phenomenon needs stochastic dynamical systems-based formulation and user-defined technique to capture the dynamics of the underlying phenomenon [38,40]. However, Industry 4.0 or smart manufacturing is perhaps in its infancy. A great deal of research lies ahead. For example, what the best architecture of new-generation manufacturing systems should be is not known yet. Some authors consider that network-based architecture is suitable for smart manufacturing [41]. Other authors consider that hierarchical architecture is suitable for smart manufacturing [42]. Some authors consider that a shift to system modeling-centric activities from (current) document meeting-centric activities is needed for smart manufacturing from both syntax [43] and semantic [44] viewpoints. Some authors consider that manufacturing decision-relevant data, information, and knowledge must be organized using bio-inspired computational frameworks for machine learning [40–45]. From the human learning viewpoint, on the other hand, manufacturing engineering-relevant educational contents can be organized in ABET-centric means [46] or other e-learning-centric means [47,48]. No matter the intended use (machine learning or human learning), or the level of sophistication (data, information, model, knowledge, simulation, and digital twin), the manufacturing engineering-relevant contents must be represented by concept maps. The reason is as follows.

Consider the evolution of web technology [49,50], as schematically illustrated in Figure 1. As seen in Figure 1, the evolution of web technology entails two dimensions. One of the dimensions is socialization, and the other is semantics (meaning of the contents). At the initial stage of web technology (Web 1.0), the communication was mainly one-way, where the users could only read the contents available on the web through the Internet. The next era (Web 2.0) materialized two-way communication, adding a functionality called writing on top of reading. This era dominates current web-based practices. The degree of socialization and semantics in these two eras has been limited. In order to increase this degree, the semantics of the contents must be increased. This leads to a concept called the semantic web [49]. Centering this new technology (semantic web), the web has been transforming into a new era called Web 3.0/4.0 [50]. The ultimate goal is to materialize personal intelligent devices. To achieve this, the contents originated in different information silos (e.g., big-data coming from different domains) must be integrated. The integration requires "meaning base" of a piece of content on top of the content itself. As a result, ontology-driven data structures of content preparation for the semantic web has evolved [51]. Thus, new-generation manufacturing is moving toward an era where the content preparation points out a scenario, as shown in Figure 2. As seen in Figure 2, in the era of Web 3.0/4.0, the manufacturing contents (i.e., different types of experimental and sensor data, models of products, machine tools, cutting tools, manufacturing processes, scheduling, and models of different manufacturing phenomena) will incorporate both the contents (syntax) and their meaning (semantics). The current trend shows that a user-defined description of the meaning of the content must be incorporated along with the content itself [25,26,30,40,45,52]. The description is inclined more toward user-defined linguistic expressions (soft) than toward predefined ontology (hard). This softness in expressing semantics leads to concept map-oriented content preparation. This means that for new-generation manufacturing, concept mapping [19–21,32] becomes a default choice, no matter the content type of contents (model, data, knowledge, and alike) [25,26,30,40].

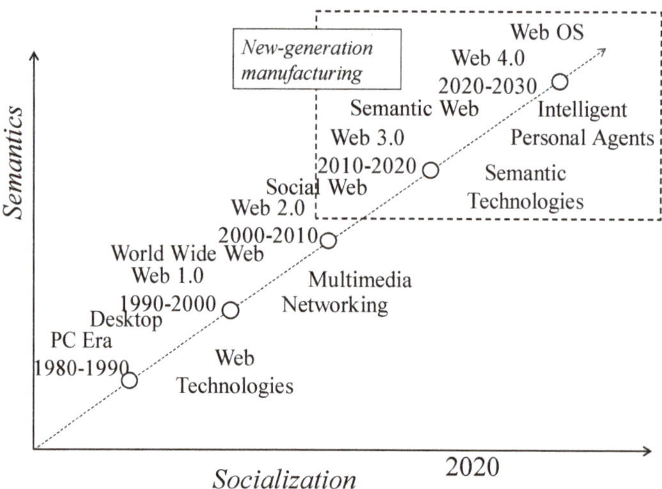


**Figure 1.** Evolution of web and new-generation manufacturing.

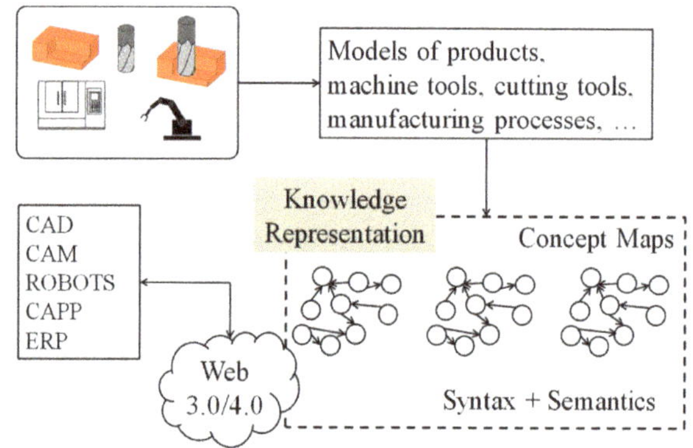


**Figure 2.** Manufacturing knowledge representation in the era of semantic web.

Apart from the issue of concept map-based content preparation for new-generation manufacturing, it (new-generation manufacturing) entails some other fundamental issues because the systems involved in new-generation manufacturing are supposed to operate as open systems. As such, independent domains might be linked as an ad hoc basis. As a result, the contents originated at unknown sources can be shared for reuse. At the same time, the contents might be altered without prior notification. This incorporates one more issue, called pragmatics [45], with syntax and semantics, as schematically illustrated in Figure 3. The contents relevant to syntax mean codified contents for machine-to-machine (M2M) communication; the contents relevant to semantics mean what does the content mean to the stakeholders. The contents relevant to pragmatics, on the other hand, must ensure that whether or not the contents are trustworthy to put into action. For the sake of better understanding, the concept of pragmatics is shown along with syntax, which are schematically illustrated in Figure 4 using an arbitrary example (i.e., the example of flower). Thus, manufacturing system developers must consider these three aspects while preparing the contents using semantic web-embedded concept maps. Now, one of the ways to tackle pragmatics, along with syntax and semantics, is to rely on the epistemic nature of the contents. This leads to a knowledge-type-aware concept map construction process. Before describing the general process of a knowledge-type-aware concept map construction process, the epistemic classification of knowledge and their origin must be articulated. This issue is described in the following section.

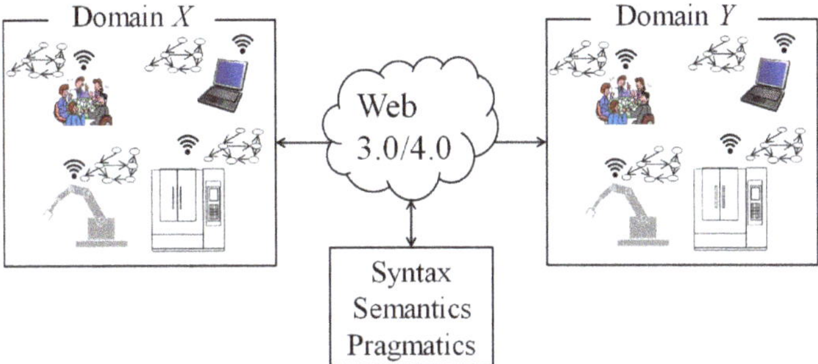


**Figure 3.** Fundamental aspects of content sharing in new-generation manufacturing.

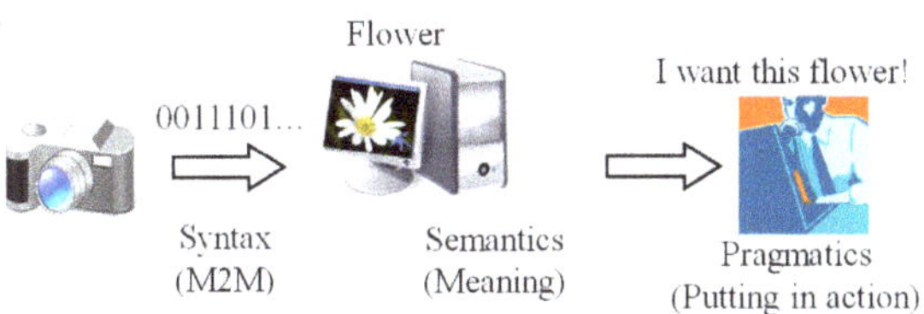


**Figure 4.** The concept of pragmatics, along with syntax and semantics.

## 3. Epistemic Classification of Knowledge

In general, a piece of knowledge is a proposition that is "justified true belief". Notably, a piece of knowledge means facts, principles, theories, and practices that are accumulated by learning; both cognitive reflections and direct experiences of individuals or groups can contribute to articulate a piece of knowledge [26]. Numerous authors have classified knowledge in different ways. From the viewpoint of an organization, knowledge is categorized into two categories, explicit knowledge and tacit knowledge [53,54]. Explicit knowledge can be formally represented for reuse, whereas tacit knowledge cannot be represented explicitly, because it is in possession of an individual or a group of individuals working in an organization (e.g., the skill to operate a machine tool). Based on semantic gravidity, knowledge can be classified into four types, namely, novice knowledge, theoretical knowledge, practical knowledge, and professional knowledge [21]. However, the author believes that epistemology (the branch of philosophy that deals with the origin and nature of knowledge) [55] is perhaps the best inspiration for classifying knowledge. Immanuel Kant [56] proposed one of the well-recognized epistemic classifications of knowledge. Kant considered both idealistic and realistic arguments of knowledge formation and proposed that there are three main types of knowledge, namely, analytic a priori, synthetic a priori, and synthetic a posteriori knowledge, as schematically illustrated in Figure 5. Analytic a priori knowledge means the knowledge gained by defining things (e.g., all bachelors are unmarried males; a triangle has three sides; feed rate is the velocity at which a cutting tool is advanced against the workpiece; and alike). Thus, analytic a priori knowledge is always true (i.e., tautology). On the other hand, synthetic a priori knowledge is the knowledge that is gained by using a mathematical deduction. Some of the examples of synthetic a priori knowledge are as follows: 7 + 4 = 11; *p* = the summation of included angles of a triangle is equal to 180°; the theoretical maximum surface roughness height of a turned surface is given as $R_t = 125\left(f^2/r_\epsilon\right)$ µm, where *f* is the feed rate (mm/rev) and $r_\epsilon$ nose radius (mm) of the cutting tool; and alike. Synthetic a priori knowledge is true within the relevant context. For example, *p* is true when the triangle is drawn on a planner surface, not on a curved surface. The same argument is valid for $R_t = 125\left(f^2/r_\epsilon\right)$ µm, i.e., it is true only when the relevant assumptions are obeyed. The other Kantian category of knowledge is called synthetic a posteriori knowledge. This category of knowledge evolves from real-world experience (or experimentation) (e.g., bachelors are rich; an apple is good for health; the feed force is less than the cutting force; and alike). As a result, synthetic a posteriori knowledge can be proven true, partially true, partially false, and even false, i.e., it is true for a stakeholder but may not necessarily be true for others (i.e., it is a matter of fact).

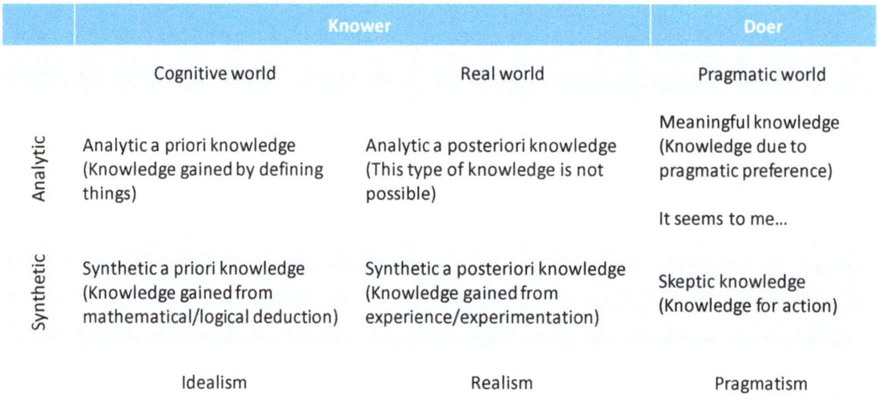

| | Knower | | Doer |
|---|---|---|---|
| | Cognitive world | Real world | Pragmatic world |
| Analytic | Analytic a priori knowledge (Knowledge gained by defining things) | Analytic a posteriori knowledge (This type of knowledge is not possible) | Meaningful knowledge (Knowledge due to pragmatic preference)<br>It seems to me... |
| Synthetic | Synthetic a priori knowledge (Knowledge gained from mathematical/logical deduction) | Synthetic a posteriori knowledge (Knowledge gained from experience/experimentation) | Skeptic knowledge (Knowledge for action) |
| | Idealism | Realism | Pragmatism |

**Figure 5.** A knowledge continuum.

However, Kantian knowledge is based on the fact that a human being is primarily a knower, as shown in Figure 5. As mentioned in the first section of this article, a human being can be a knower and doer at the same time. In the doer mode, a human being remains pragmatic and gives priority to her/his preferences and judgment, even though everything is not clearly known in terms of other forms of knowledge. The knowledge gained in pragmatic mode can be of two types. One of the types evolves in the analytic mode. It is referred to as meaningful knowledge as if it is the outcome of meaningful learning. It thus injects new concepts, as it is the result of an individual's preference: "It seems to me that . . . ." For example, consider the following proposition: "A cutting tool having an oval cross-sectional area performs better than a cutting tool having a circular cross-sectional area while removing material around sharps corner of a workpiece in milling." This is an outcome of meaningful learning because it injects a new cutting tool (oval-shaped tool) to solve a problem (remove material in the sharp corners of a workpiece). This proposition cannot be proven true or false until there is a piece of supportive synthetic a priori or synthetic a posteriori knowledge available. As far as manufacturing is concerned, this type of knowledge can dominate other types of knowledge. The other type of knowledge in doer mode is defined as skeptic knowledge. It evolves in the synthetic mode. It may or may not inject new concepts. It is directly related to other types of knowledge. Skeptic knowledge often leads learners to ponder a course of action to study further. Thus, it is somehow linked to other types of knowledge; that is, it is a purposeful interpretation of an individual based on other available pieces of knowledge. For example, consider the following proposition: "Reduce feed rate to ensure a better surface finish." It is a piece of skeptic knowledge and helps take a course of action (e.g., optimize a material removal process), even though the rationale is somewhat informal. However, when skeptic knowledge is directly related to a piece of meaningful knowledge, the skepticism regarding it (meaningful knowledge) manifests the skepticism regarding skeptic knowledge. At the same time, it follows other concept maps as a part of further study regarding the subject matter. It means that skeptic knowledge acts as a tool for enhancing meaningful learning among learners. This is exemplified in the next section.

Nevertheless, other than the analytic a priori knowledge, all the types of knowledge mentioned above can be proven false. This means that a learner, either a human being or a machine, can set a strategy to verify or validate whether or not a given piece of knowledge is true before using it. The strategy will depend on the type of knowledge. For example, if the type of knowledge is synthetic a priori, the learner is supposed to find out the deductive steps and relevant definitions (analytic a priori knowledge) to determine the truthfulness of it. If the type of knowledge is synthetic a posteriori, the learner is supposed to find out the rationales and integrity of the relevant experimental results or experience to determine the truthfulness of it. If the type of knowledge is meaningful knowledge, then the learner must identify the innovative process that leads to the conclusion (meaningful knowledge) or identify the analytical or experimental processes that lead to the meaningful knowledge. If the type of knowledge is skeptic knowledge, then the learner must extract the pieces of relevant pieces of background knowledge (analytical a priori, systematic a priori, synthetic a posteriori, or meaningful knowledge) that helped the knowledge formulator to conceive the skeptic knowledge for taking actions.

## 4. Concept Map Creation and Analysis

Having described the fundamental issues and types of knowledge, it is time to create and analyze a concept map containing manufacturing engineering contents.

It is worth mentioning that many authors have studied subject matter-based educational needs and relevant ICT infrastructures from the viewpoint of new-generation manufacturing [25,46,57–60]. Some authors have emphasized a particular type of knowledge structure and its transformation [61] for the sake of active learning and teaching. Some authors have emphasized concept map-based content preparation [25] for enhancing meaningful learning in manufacturing [16–21]. In addition, many concept maps carrying both theoretical and empirical knowledge of manufacturing can be found in [25,30,31,40]. Some of them are for human learning [25], and some of them are for machine learning [30,31].

However, the concept map-based learning performances of some undergraduate engineering students have been reported in [25]. One of the remarks made by the learners is about the size of the concept map (large size makes the content less attractive and cumbersome). Another important observation made by the learners is that some of the concepts are controversial, and, thereby, difficult to comprehend, even though excellent illustrations and references are embedded in the concept maps. As a result, the central theme of meaningful learning (avoiding memorization [16–21]) gets affected. The root cause of this is perhaps the presence of synthetic a posteriori knowledge, meaningful knowledge, and skeptic knowledge in the concept map. This means that some of the concepts may appear to be analytic a priori, but they are meaningful knowledge or even skeptic knowledge. In this case, the instructor may put more effort into explaining these pieces of knowledge so that the learners avoid memorization or seek other reference materials to grasp the real meaning or counterexamples. As a result, knowledge-aware concept mapping can make the maps systematic and comprehensible to learners. At the same time, the instructor can carry out the teaching activities in a more systematic manner.

For example, a relatively small concept map is constructed based on the concept map of a turning (a widely used manufacturing process) shown in [25]. For the sake of better understanding, a well-known subject matter—the shear plane theory of a material removal process—is considered, as illustrated in Figure 6. Mechanical or manufacturing engineering students learn the shear plane theory (Figure 6) at the undergraduate level, which is an oversimplified model of chip formation in material removal processes [62]. The symbols shown in Figure 6 have their usual meaning. Numerous propositions can represent the knowledge underlying the contents shown in Figure 6. Each proposition should be based on a type of knowledge described in the previous section. The number of propositions depends on the individual who formulates those.

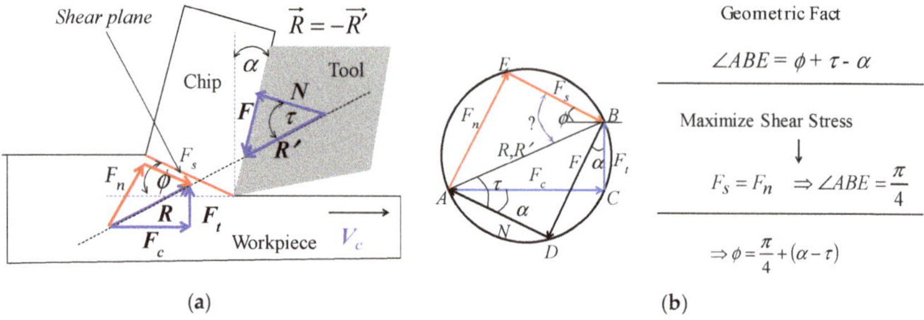


**Figure 6.** Shear plane theory of material removal process. (**a**) Shear plane theory; (**b**) position of shear plane angle.

Figure 7 shows a screen-print of the semantic web constructed to represent the knowledge underlying Figure 6. It can be accessed through the Internet from the URL shown in [63]. The icons shown in the nodes called "here" have links to the illustrations shown in Figure 6. If needed, other contents (video clip and other concept maps) can be linked to the appropriate nodes of the concept map shown in Figure 7. Since there are no experimental facts associated with Figure 7, there are no propositions regarding the synthesis a posteriori knowledge in Figure 7. Synthetic a posteriori knowledge relevant to turning can be found in the concept maps reported in [25,30,45].

The concept map shown in Figure 7 boils down to the following propositions.

(1) The components of two balancing forces (illustrated in Figure 6) act while removing materials in the form of a chip from a workpiece.
(2) Workpiece materials become the chip from a plane called shear plane due to the action of a cutting tool.
(3) The chip forms by maximizing the shear force, resulting in the included angle ∠ABE (illustrated in Figure 6) equal to $\pi/4$.
(4) Shear force acts along the shear plane.
(5) The included angle ∠ABE is equal to $\phi + \tau - \alpha$ based on the geometric relationships among the three pairs of forces illustrated in Figure 6.
(6) Two balancing forces (illustrated in Figure 6) can be represented by three pairs of forces, where each pair consists of two orthogonal components.
(7) Cutting force acts in the direction of cutting velocity.
(8) Cutting force is a component of one of the three pairs of forces (illustrated in Figure 6).
(9) The shear plane angle is given as $\phi = \pi/4 + (\alpha - \tau)$.

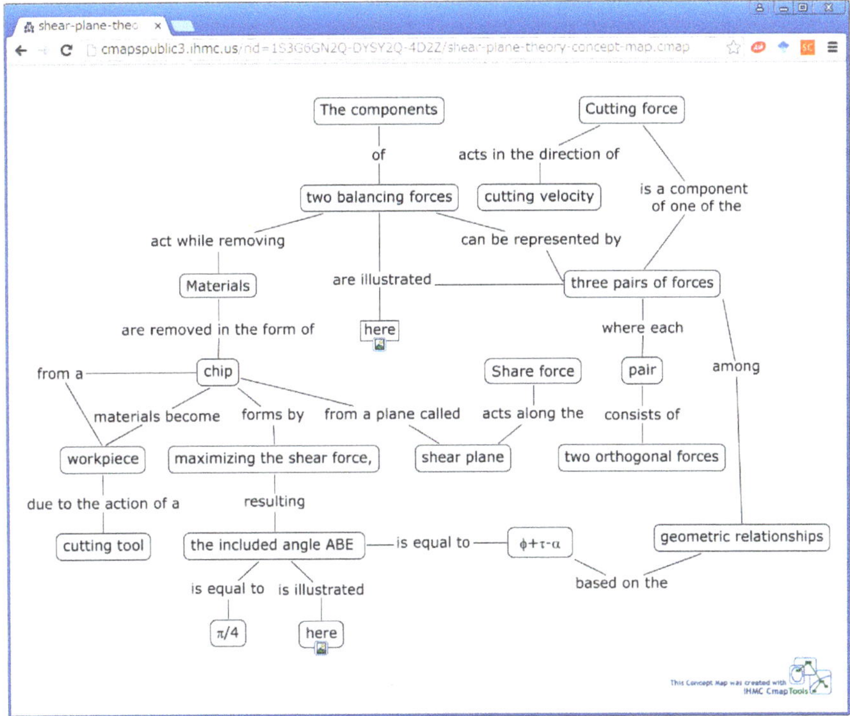


**Figure 7.** Print-screen of a semantic web-embedded concept map.

Consider the first proposition: The components of two balancing forces (illustrated in Figure 6) act while removing materials in the form of the chip from a workpiece. It is a piece of synthetic a priori knowledge because it comes from a mathematical deduction that two forces must act along a line but in the opposite directions to maintain a balance. This is true because the external forces acting on the workpiece makes a static balance; otherwise, the workpiece may move out from the holding devices.

Consider the second statement: Workpiece materials become the chip from a plane called shear plane due to the action of a cutting tool. This proposition is a sophisticated form of knowledge because it entails two forms of knowledge, namely, analytic a priori knowledge and meaningful knowledge. If the proposition is rewritten using two propositions, then this duality of knowledge can be understood. For example, consider the following two propositions: (2-1) Workpiece materials become the chip due to the action of a cutting tool; and (2-2) Workpiece materials become the chip from a plane called shear plane. Proposition 2-1 is a piece of analytic a priori knowledge because it just defines the manufacturing process (chip formation is necessary for achieving material removal in machining), as well the role of a cutting tool. On the other hand, the other proposition (proposition 2-2) is a piece of meaningful knowledge. The reason is that one has considered this concept (shear plane) to explain chip formation. Chip formation can occur from a region (not a plane) [62]. This means that the idea of the shear plane is somewhat controversial at this point. As far the knowledge-aware construction of concept maps is concerned, the map shown in Figure 7 needs revision, avoiding the mixing of the types of knowledge.

Consider the third proposition: The chip forms by maximizing the shear force, resulting in the included angle ∠ABE (illustrated in Figure 6) equal to $\pi/4$. It is also a piece of meaningful knowledge because a new concept is injected (maximizing the shear force) to set the value of the angle ∠ABE. There are other ways to perceive this case. For example, one can consider instead that "minimizing energy" is the phenomenon that takes place while removing the chip from a workpiece during machining [62].

Therefore, comprehending this knowledge requires some effort from the learner's side and making it compressible requires efforts from the instructor's side.

Consider the fourth proposition: Shear force acts along the shear plane. It is a piece of analytic a priori knowledge that defines the concept of shear force with respect to shear plane. Even though this proposition is a piece of analytic a priori knowledge, it is coupled with another concept (shear plane) that also needs to be defined. Since the concept of shear plane refers to a piece of meaningful knowledge, the comprehension regarding the shear force creates an amount of fuzziness among the learners. Therefore, this piece of knowledge creates what can be referred to as fuzzy circularity. Therefore, both concepts (shear force and shear plane) must be handled with care during the process of learning and teaching.

Consider the fifth proposition: The included angle $\angle$ABE is equal to $\phi + \tau - \alpha$ based on the geometric relationships among the three pairs of forces illustrated in Figure 6. It is a piece of synthetic a priori knowledge because $\angle$ABE = $\phi + \tau - \alpha$ is deduced from the geometric relationships shown in Figure 6. Therefore, if the learns can follow the steps used in the deduction, the validity of the proposition becomes clear to them. The instructor may set some exercises for the learners to master the steps.

Consider the sixth proposition: Two balancing forces (illustrated in Figure 6) can be represented by three pairs of forces, where each pair consists of two orthogonal components. It is a piece of synthetic a priori knowledge because, from a mathematical point of view, a planner force can be deduced into two orthogonal components. Regarding the three pairs of forces, the following comments can be made. Consider, for example, the force acting between the cutting tool surface and the chip (Figure 6). Whenever an object slides against another, friction force occurs, and its magnitude depends on the surface condition (coefficient of friction) and the force acting normal to the sliding surfaces. This kind of explanation is deductive truth that relates basic knowledge of physics and engineering science. The same arguments hold for the other two pairs of forces. Therefore, there is no problem treating proposition 6 as a piece of synthetic a priori knowledge.

Consider the seventh proposition: Cutting force acts in the direction of cutting velocity. It is a piece of analytic a priori knowledge because it defines cutting force; that is, a force that acts in the direction of cutting velocity. Even though this proposition is a piece of analytic a priori knowledge, it is coupled with another concept (cutting velocity) that also needs to be defined. Therefore, this piece of knowledge creates a circularity. This time it does not entail fuzzy circularity, unlike the case for proposition 4. This time, it is rather a simple circularity. Nevertheless, both concepts (cutting force and cutting velocity) deserve explanation with respect to each other, requiring extra care from both the learner's and instructor's sides.

Consider the eighth proposition: Cutting force is a component of one of the three pairs of forces (illustrated in Figure 6). It is a piece of synthetic a priori knowledge because, from the knowledge of mathematics (vector algebra), it is clear that a planner force can be decomposed into orthogonal components.

Consider the last proposition: The shear plane angle is given as $\phi = \pi/4 + (\alpha - \tau)$. It is a piece of skeptic knowledge, though it seems a piece of synthetic a priori knowledge. The reasons are two-fold. The first reason is its epistemic nature, and the other reason is its ability to trigger other learning activities that can incorporate other concept maps. This is schematically illustrated in Figure 8. As seen in Figure 8, at least two new concept maps, denoted as Map-1 and Map-2, can evolve due to this piece of skeptic knowledge. While pursuing Map-1, the link of the proposition 9 (skeptic knowledge) with other propositions can be considered. It is directly related to propositions 2, 3, and 4, which entail meaningful knowledge. Since a great deal of skepticism is already associated with propositions 2, 3, and 4, as described above, proposition 9 is automatically subjected to a great deal of skepticism. Therefore, learners can seek other pieces of meaningful knowledge (say, "the chip forms by minimizing energy at the shear please") [62]. In this case, the included angle $\angle$ABE (illustrated in Figure 6) will no longer be equal to $\pi/4$, resulting in a relationship other than $\phi = \pi/4 + (\alpha - \tau)$. If so, new pieces of knowledge (say,

$\phi = \pi/4 + 0.5(\alpha - \tau)$) may evolve [62]. This results in a new concept map, denoted as Map-1. Consider the concept map Map-2. The learning activity can also be directed toward incorporating some pieces of synthetic a posteriori knowledge (experimental facts). In this case, the apparent learning activities are as follows. According to Figure 6, the ratio between the thicknesses of the undeformed material before chip formation and deformed material after chip formation (denoted as $r$) is equal to $\sin(\phi)/\cos(\phi - \alpha)$. An experiment can be carried out to know $r$ for a predefined $\alpha$ (rake angle). If these experimentally determined values of $r$ and $\phi$ are input in $r = \sin(\phi)/\cos(\phi - \alpha)$, the value of $\phi$ can be calculated. This calculated value can be compared to the theoretical one, $\phi = \pi/4 + (\alpha - \tau)$. Thus, proposition 9 (skeptic knowledge) leads to some learning activities to know about the nature of machining from different points of view, to see whether or not the associated meaningful knowledge can be trusted. In other words, proposition 9 can enhance discipline-based education.

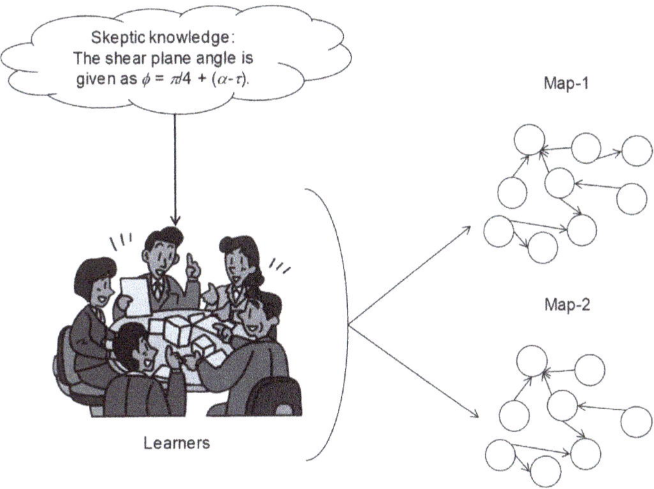


**Figure 8.** Learning enhancement through skeptic knowledge.

## 5. Conclusions

Though concept mapping has contributed to different levels of education, its construction and deployment processes require a great deal of study. As far as manufacturing engineering is concerned, both learning factory and human-cyber-physical systems (i.e., human and machining learning) have opened new opportunities for concept maps and brought new challenges as well. The proposed types of knowledge (analytic a priori, synthetic a priori, synthetic a posteriori, meaningful, and skeptic knowledge) can help exploit the abovementioned opportunities and tackle the challenges. This is clear from the contents presented in the previous few sections.

In order to educate students with the basic knowledge of material removal (a common topic that all manufacturing engineering students study at undergraduate degree level), a nine proposition-based concept map has been introduced. The map represents knowledge of the shear plane theory of material removal. All types of knowledge are integrated into the map, except synthetic a posteriori knowledge. The remarkable thing is that the learning process directs the learners to form new concept maps, wherein the not-yet-included type of knowledge (in this case, synthetic a posteriori knowledge) is likely to appear. This directs the whole learning process in the direction of meaningful learning.

Instead of the presented concept map, a concept map without embedding synthetic a priori knowledge can be considered for the same purpose (educating students with the basic knowledge of material removal) and can be observed whether or not the learners propose another concept map that incorporates synthetic a priori knowledge. This means that there is enough room for further investigations showing the education capability of concept maps and, thereby, meaningful learning, while offering manufacturing engineering education.

**Funding:** This research received materials and financial supports from Kitami Institute of Technology.

**Conflicts of Interest:** The author declares no conflicts of interest.

## References

1. Wilson, S.; Schweingruber, H.; Nielsen, N. (Eds.) *Science Teachers' Learning: Enhancing Opportunities, Creating Supportive Contexts*; The National Academies Press: Washington, DC, USA, 2015. [CrossRef]
2. Bamberger, Y.M.; Davis, E.A. Middle school science students' scientific modelling performances across content areas and within a learning progression. *Int. J. Sci. Educ.* **2013**, *35*, 213–238. [CrossRef]
3. Berland, L.K.; Reiser, B.J. Making sense of argumentation and explanation. *Sci. Educ.* **2009**, *93*, 26–55. [CrossRef]
4. McNeill, K.L. Elementary students' views of explanation, argumentation and evidence and abilities to construct arguments over the school year. *J. Res. Sci. Teach.* **2011**, *48*, 793–823. [CrossRef]
5. McNeill, K.L.; Lizotte, D.J.; Krajcik, J.; Marx, R.W. Supporting students' construction of scientific explanations by fading scaffolds in instructional materials. *J. Learn. Sci.* **2006**, *15*, 153–191. [CrossRef]
6. Lehrer, R.; Schauble, L. Seeding evolutionary thinking by engaging children in modeling its foundations. *Sci. Educ.* **2012**, *96*, 701–724. [CrossRef]
7. Smith, C.L.; Wiser, M.; Anderson, C.W.; Krajcik, J. Implications of research on children's learning for standards and assessment: A proposed learning progression for matter and the atomic molecular theory. *Meas. Interdiscip. Res. Perspect.* **2006**, *4*, 1–98. [CrossRef]
8. National Research Council. *How People Learn: Bridging Research and Practice*; Donovan, M.S., Bransford, J.D., Pellegrino, J.W., Eds.; National Academy Press: Washington, DC, USA, 1999.
9. National Research Council. *A Framework for K-12 Science Education Practices, Crosscutting Concepts, and Core Ideas*; The National Academies Press: Washington, DC, USA, 2012.
10. Chi, M.T.H. Commonsense conceptions of emergent processes: Why some misconceptions are robust. *J. Sci.* **2005**, *14*, 161–199. [CrossRef]
11. Chi, M.T.H. Three types of conceptual change: Belief revision, mental model transformation, and categorical shift. In *International Handbook of Research on Conceptual Change*; Vosniadou, S., Ed.; Routledge: New York, NY, USA, 2008; pp. 61–82.
12. Singer, S.R.; Nielsen, N.R.; Schweingruber, H.A. (Eds.) *Discipline-Based Education Research: Understanding and Improving Learning in Undergraduate Science and Engineering*; The National Academies Press: Washington, DC, USA, 2012. [CrossRef]
13. Criteria for Accrediting Engineering Programs, 2019–2020. Available online: https://www.abet.org/accreditation/accreditation-criteria/criteria-for-accrediting-engineering-programs-2019-2020/#GC2 (accessed on 23 July 2019).
14. Brown, A.L. Knowing when, where, and how to remember: A problem of metacognition. In *Advances in Instructional Psychology*; Glaser, R., Ed.; Lawrence Erlbaum Associates: Hillsdale, NJ, USA, 1978; Volume 2, pp. 77–165.
15. Flavell, J.H. Metacognition and cognitive monitoring: A new area of cognitive-developmental inquiry. *Am. Psychol.* **1979**, *34*, 906–911. [CrossRef]
16. Ausubel, D.P. *Educational Psychology, a Cognitive View*; Holt, Rinehart and Winston: New York, NY, USA, 1968.
17. Ausubel, D.P.; Novak, J.D.; Hanesian, H. *Educational Psychology: A Cognitive View*, 2nd ed.; Holt, Rinehart and Winston: New York, NY, USA, 1978.
18. Ausubel, D.P. *The Acquisition and Retention of Knowledge: A Cognitive View*; Kluwer Academic: Boston, MA, USA, 2000.
19. Novak, J.D.; Gowin, D.B. *Learning How to Learn*; Cambridge University Press: New York, NY, USA, 1984.
20. Novak, J.D. *Learning, Creating, and Using Knowledge: Concept Maps as Facilitative Tools in Schools and Corporations*, 2nd ed.; Rutledge: New York, NY, USA, 2012.
21. Kinchin, I.M.; Möllits, A.; Reiska, P. Uncovering Types of Knowledge in Concept Maps. *Educ. Sci.* **2019**, *9*, 131. [CrossRef]

22. Marzetta, K.; Mason, H.; Wee, B. 'Sometimes They Are Fun and Sometimes They Are Not': Concept Mapping with English Language Acquisition (ELA) and Gifted/Talented (GT) Elementary Students Learning Science and Sustainability. *Educ. Sci.* **2018**, *8*, 13. [CrossRef]
23. Young, J.; Young, J.; Cason, M.; Ortiz, N.; Foster, M.; Hamilton, C. Concept Raps versus Concept Maps: A Culturally Responsive Approach to STEM Vocabulary Development. *Educ. Sci.* **2018**, *8*, 108. [CrossRef]
24. Zhou, Y. A Concept Tree of Accounting Theory: (Re)Design for the Curriculum Development. *Educ. Sci.* **2019**, *9*, 111. [CrossRef]
25. Ullah, A.M.M.S. On the interplay of manufacturing engineering education and e-learning. *Int. J. Mech. Eng. Educ.* **2016**, *44*, 233–254. [CrossRef]
26. Abele, E.; Chryssolouris, G.; Sihn, W.; Metternich, J.; ElMaraghy, H.; Seliger, G.; Sivard, G.; ElMaraghy, W.; Hummel, V.; Tisch, M.; et al. Learning factories for future oriented research and education in manufacturing. *CIRP Ann. Manuf. Technol.* **2017**, *66*, 803–826. [CrossRef]
27. Baena, F.; Guarin, A.; Mora, J.; Sauza, J.; Retat, S. Learning Factory: The Path to Industry 4.0. *Procedia Manuf.* **2017**, *9*, 73–80. [CrossRef]
28. Tisch, M.; Hertle, C.; Abele, E.; Metternich, J.; Tenberg, R. Learning factory design: A competency-oriented approach integrating three design levels. *Int. J. Comput. Integr. Manuf.* **2016**, *29*, 1355–1375. [CrossRef]
29. Schuh, G.; Anderl, R.; Gausemeier, J.; Hompel, M.; Wahlster, W. (Eds.) *Industrie 4.0 Maturity Index. Managing the Digital Transformation of Companies (Acatech Study)*; Herbert Utz Verlag: Munich, Germany, 2017.
30. Ullah, A.M.M.S. Modeling and simulation of complex manufacturing phenomena using sensor signals from the perspective of Industry 4.0. *Adv. Eng. Inform.* **2019**, *39*, 1–13. [CrossRef]
31. Ullah, A.M.M.S.; Arai, N.; Watanabe, M. Concept Map and Internet-aided Manufacturing. *Procedia CIRP* **2013**, *12*, 378–383. [CrossRef]
32. Cañas, A.J.; Novak, J.D.; Reiska, P. How good is my concept map? Am I a good Cmapper? *Knowl. Manag. E-Learn.* **2015**, *7*, 6–19.
33. Zhao, M.; Wang, H.; Guo, J.; Liu, D.; Xie, C.; Liu, Q.; Cheng, Z. Construction of an Industrial Knowledge Graph for Unstructured Chinese Text Learning. *Appl. Sci.* **2019**, *9*, 2720. [CrossRef]
34. Zheng, P.; Wang, H.; Sang, Z.; Zhong, R.Y.; Liu, Y.; Liu, C.; Mubarok, K.; Yu, S.; Xu, X. Smart manufacturing systems for Industry 4.0: Conceptual framework, scenarios, and future perspectives. *Front. Mech. Eng.* **2018**, *13*, 137–150. [CrossRef]
35. Kusiak, A. Smart manufacturing. *Int. J. Prod. Res.* **2018**, *56*, 508–517. [CrossRef]
36. Lasi, H.; Fettke, P.; Kemper, H.G.; Feld, T.; Hoffmann, M. Industry 4.0. *Bus. Inf. Syst. Eng.* **2014**, *6*, 239–242. [CrossRef]
37. Koren, Y.; Gu, X.; Guo, W. Reconfigurable manufacturing systems: Principles, design, and future trends. *Front. Mech. Eng.* **2018**, *13*, 121–136. [CrossRef]
38. Ullah, A.M.M.S. Surface Roughness Modeling Using Q-Sequence. *Math. Comput. Appl.* **2017**, *22*, 33. [CrossRef]
39. Imran Shafiq, S.; Szczerbicki, E.; Sanin, C. Decisional-DNA Based Smart Production Performance Analysis Model. *Cybern. Syst.* **2019**, *50*, 154–164. [CrossRef]
40. Ghosh, A.K.; Ullah, A.M.M.S.; Kubo, A. Hidden Markov Model-Based Digital Twin Construction for Futuristic Manufacturing Systems. *Artif. Intell. Eng. Des. Anal. Manuf.* **2019**, *33*, 317–331. [CrossRef]
41. Vogel-Heuser, B.; Kegel, G.; Bender, K.; Wucherer, K. Global Information Architecture for Industrial Automation. *Autom. Prax.* **2009**, *51*, 108–115. [CrossRef]
42. Lee, J.; Bagheri, B.; Kao, H.A. A cyber-physical systems architecture for Industry 4.0-based manufacturing systems. *Manuf. Lett.* **2015**, *3*, 18–23. [CrossRef]
43. Schroeder, G.N.; Steinmetz, C.; Pereira, C.E.; Espindola, D.B. Digital Twin Data Modeling with AutomationML and a Communication Methodology for Data Exchange. *IFAC-PapersOnLine* **2016**, *49*, 12–17. [CrossRef]
44. Madni, A.M.; Madni, C.C.; Lucero, S.D. Leveraging Digital Twin Technology in Model-Based Systems Engineering. *Systems* **2019**, *7*, 7. [CrossRef]
45. Ullah, A.M.M.S.; D'Addona, D.; Arai, A. DNA based computing for understanding complex shapes. *Biosystems* **2014**, *117*, 40–53. [CrossRef] [PubMed]
46. Ullah, A.S.; Harib, K.H. Tutorials for Integrating CAD/CAM in Engineering Curricula. *Educ. Sci.* **2018**, *8*, 151. [CrossRef]

47. Salah, B.; Abidi, M.H.; Mian, S.H.; Krid, M.; Alkhalefah, H.; Abdo, A. Virtual Reality-Based Engineering Education to Enhance Manufacturing Sustainability in Industry 4.0. *Sustainability* **2019**, *11*, 1477. [CrossRef]
48. Chong, S.; Pan, G.T.; Chin, J.; Show, P.L.; Yang, T.C.K.; Huang, C.M. Integration of 3D Printing and Industry 4.0 into Engineering Teaching. *Sustainability* **2018**, *10*, 3960. [CrossRef]
49. Berners-Lee, T.; Hendler, J.; Lassila, O. The semantic web. *Sci. Am.* **2001**, *284*, 34–43. [CrossRef]
50. Pileggi, S.F.; Fernandez-Llatas, C.; Traver, V. When the Social Meets the Semantic: Social Semantic Web or Web 2.5. *Future Internet* **2012**, *4*, 852–864. [CrossRef]
51. Khilwani, N.; Harding, J.A.; Choudhary, A.K. Semantic web in manufacturing. *Proc. Inst. Mech. Eng. Part B J. Eng. Manuf.* **2009**, *223*, 905–925. [CrossRef]
52. Alkahtani, M.; Choudhary, A.; De, A.; Harding, J.A. A decision support system based on ontology and data mining to improve design using warranty data. *Comput. Ind. Eng.* **2019**, *128*, 1027–1039. [CrossRef]
53. Polanyi, M. *Personal Knowledge: Towards a Post-Critical Philosophy*; The University of Chicago Press: Chicago, IL, USA, 1958.
54. Nonaka, I. A Dynamic Theory of Organizational Knowledge Creation. *Organ. Sci.* **1994**, *5*, 14–37. [CrossRef]
55. Sosa, E. *Epistemology*; Princeton University Press: Princeton, NJ, USA, 2017.
56. Kant, I. *Critique of Pure Reason: The Cambridge Edition of the Works of Immanuel Kant*; Guyer, P., Wood, A.W., Eds.; Cambridge University Press: Cambridge, UK, 2000.
57. Tvenge, N.; Martinsen, K. Integration of digital learning in industry 4.0. *Procedia Manuf.* **2018**, *23*, 261–266. [CrossRef]
58. Umachandran, D.K.; Jurcic, I.; Ferdinand-James, D.; Said, M.; Rashid, A. Gearing Up Education Towards Industry 4.0. *IJCT* **2018**, *17*, 7305–7311. [CrossRef]
59. Chou, P.N.; Feng, S.T. Using a Tablet Computer Application to Advance High School Students' Laboratory Learning Experiences: A Focus on Electrical Engineering Education. *Sustainability* **2019**, *11*, 381. [CrossRef]
60. Coşkun, S.; Kayıkcı, Y.; Gençay, E. Adapting Engineering Education to Industry 4.0 Vision. *Technologies* **2019**, *7*, 10. [CrossRef]
61. Kolb, D. *Experiential Learning as the Science of Learning and Development*; Prentice Hall: Englewood Cliffs, NJ, USA, 1984.
62. Astakhov, V.P. On the inadequacy of the single-shear plane model of chip formation. *Int. J. Mech. Sci.* **2005**, *47*, 1649–1672. [CrossRef]
63. The Semantic Web. Available online: http://cmapspublic3.ihmc.us/rid=1S3G6GN2Q-DYSY2Q-4D2Z/shear-plane-theory-concept-map.cmap (accessed on 23 July 2019).

MDPI
St. Alban-Anlage 66
4052 Basel
Switzerland
Tel. +41 61 683 77 34
Fax +41 61 302 89 18
www.mdpi.com

*Education Sciences* Editorial Office
E-mail: education@mdpi.com
www.mdpi.com/journal/education